STERLING
Education

Essential Chemistry

Chemical Bonding

4th edition

STERLING
Education

From the foundations of chemical reactions to the complex mechanisms of atomic particles, *Essential Chemistry Self-Teaching Guides* are a comprehensive compendium of clearly explained texts to learn and master these multifaceted chemistry topics.

These guides provide a detailed review of the fundamental mechanisms of chemical and physical processes at the atomic level. Develop a better understanding of the electronic structure of elements, principles of chemical bonding, phases of matter, types and mechanisms of chemical reactions, and principles of solutions and acid-base equilibria. Learn about rate processes in chemical reactions, empirical and molecular formulas, enthalpy, entropy, oxidation number, the laws of thermodynamics, and electrochemistry. Reinforce your learning by working through the practice questions and step-by-step solutions.

Created by highly qualified chemistry instructors, researchers, and education specialists, these books empower readers by helping them increase their understanding of general chemistry.

We sincerely hope that these guides are valuable for your learning.

250909akp

Featured on

Electronic Structure & Periodic Table

Chemical Bonding

States of Matter & Phase Equilibria

Stoichiometry

Solution Chemistry

Chemical Kinetics & Equilibrium

Acids & Bases

Chemical Thermodynamics

Electrochemistry

Visit our Amazon store

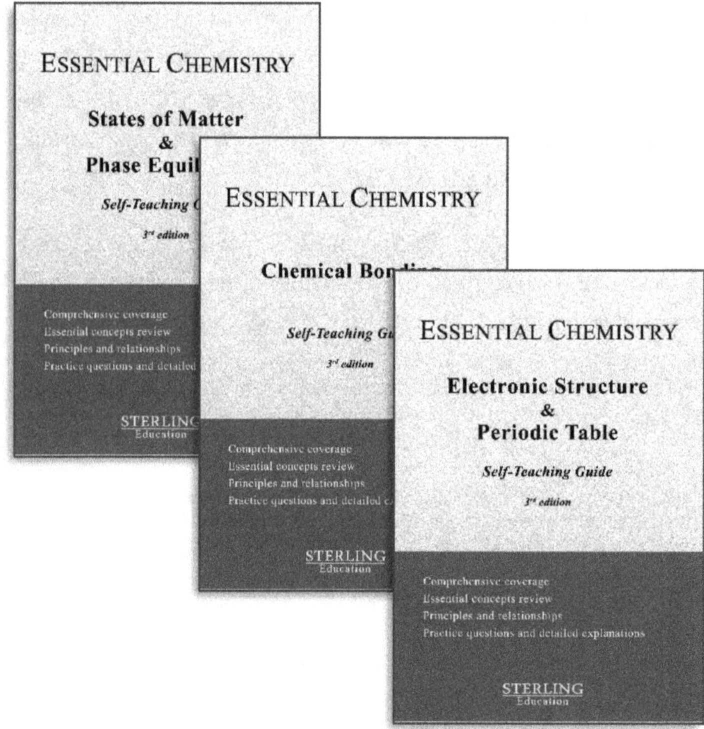

Essential Physics Self-Teaching Guides

Kinematics and Dynamics

Equilibrium and Momentum

Force, Motion and Gravitation

Work and Energy

Fluids and Solids

Waves and Periodic Motion

Light and Optics

Sound

Electrostatics and Electromagnetism

Electric Circuits

Thermal Physics

Atomic and Nuclear Physics

Visit our Amazon store

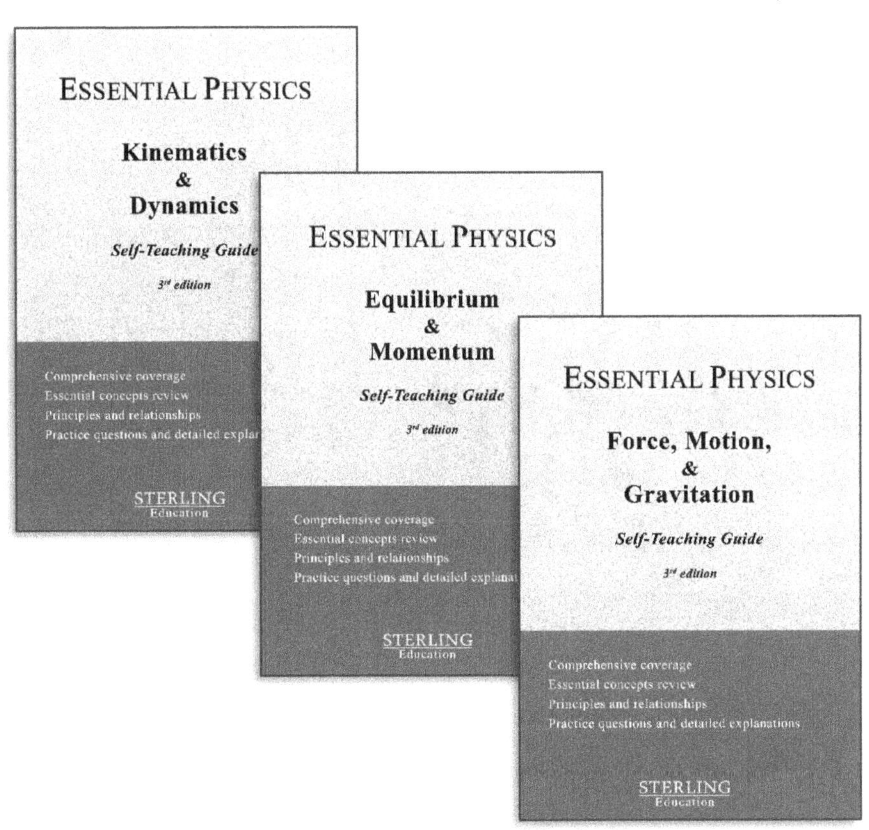

Everything You Always Wanted to Know About...

Chemistry

Physics

Cell & Molecular Biology

Organismal Biology

Human Anatomy & Physiology

Environmental Science

American History

American Law

American Government & Politics

Comparative Government & Politics

World History

European History

Psychology

Sociology

Human Geography

Visit our Amazon store

For online practice resources visit

https://www.sterling-prep.com

If you benefited from this book, please leave a review on Amazon so others
can learn from your input. Reviews help us understand our customers'
needs and experiences while keeping our commitment to quality.

Table of Contents

Table of Contents (*continued*)

Table of Contents (*continued*)

Page intentionally left blank

REVIEW

Chemical Bonding

The Nature of Science

Electrostatic Forces Between Ions

Electron Pair Sharing

Sigma and *Pi* Bonds

Hybridization

Valence Bond (VB) Theory

Molecular Orbital (MO) Theory

Lewis Electron Dot Formulas

Formal Charge

Resonance

Partial Ionic Character

Polarity

Page intentionally left blank

The Nature of Science

Scientific discovery

Observation is the first step toward scientific discovery. Observations recognize patterns and anomalies in unexplained phenomena.

Hypothesis is advanced as a proposed explanation for the observation.

Evaluating a hypothesis involves experiments and additional observations (i.e., recording and collecting data).

Experimental observations are made while manipulating *independent variables* (e.g., how long the plant is exposed to sunlight per twenty-four hours) with *dependent variables* (data collected regarding height measurements).

Observations generate *data* that:

1) *support the hypothesis* (i.e., different from proves) or

2) *refute the hypothesis*.

Hypothesis are modified by narrowing or expanding their application to explain additional data. More experiments are performed, and the results either *support* (not prove) or *refute* the hypothesis during this iterative process.

Principles are the initially proposed explanations that are specific and apply to a narrow range of phenomena.

Theories build upon valid hypotheses

Theories are proposed to account for the observations when a hypothesis is valid over a range (e.g., sunlight, temperature, or humidity).

Theories are advanced to explain the phenomena and to predict what will happen under other conditions with different variables.

Further observations determine whether the predictions were accurate.

Research (re-search or *again search*) continues as these many observations support hypotheses and theories.

When theories are accurate over various scenarios and experiments, *Laws* are developed based on these repeatedly replicated results.

Laws as established theories

Laws are brief descriptions of how nature behaves in a broad set of circumstances. They are the consolidation of several theories. Laws are the products of multiple theories tested and proven, culminating in a broad claim regarding those predictions.

Laws are widely applicable explanations. Laws with broad and straightforward claims are more robust than those that make claims on a narrower, more specific set of variables and parameters.

The number of Laws describing physical phenomena is minuscule compared to the number of theories and magnitudes, which are less than the number of supported hypotheses.

The number of refuted (unsupported) hypotheses may be several magnitudes larger.

Failed experiments do not typically refer to the physical process where the researcher made a human error during the experiment. A failed experiment means that the *data does not support the hypothesis.* Therefore, after additional trials to validate the "failed" results, the hypothesis must be modified to predict the results of either future experiments or objective observations in the physical world.

Theories, models and laws

Theories emerge from a hypothesis that experimental data has repeatedly supported. Theories are detailed statements that provide testable predictions of the behavior of natural phenomena.

Theories are established to explain observations and then tested (supported or refuted) based on their predictions. Tested theories can articulate the boundaries of the tested hypothesis.

Theories are more specific than *hypotheses* but vaguer and more encompassing than *models*.

Models are constructed to explain the observed behavior if a hypothesis accurately predicts a phenomenon. They are more specific in their application than theories.

Data collected to develop the theories provide the conceptual foundation for constructing models. The model creates mental pictures (e.g., complicated weather patterns displayed as visual pictures for the viewer) of the physical phenomena.

Models should be consistent with scientific theories' predictions and support a comprehensive understanding of a phenomenon. Models allow for predictive outcomes under specific theories (e.g., when the wind speed reaches x, then y will occur).

Multiple models are integrated to explain a portion, or the entire scope, of a *theory*.

Understand the limitations of a model and do not apply it dogmatically unless its applicability is evaluated. A model is an underlying representation of a part of a theory; it is not intended to provide a complete picture of every phenomenon under that theory.

Instead, models provide the fundamental aspects of the theory.

Electrostatic Forces Between Ions

Ionic bond formation

An *ionic bond* forms when electrons transfer from one atom to another. This electron transfer results in oppositely charged species that attract each other via electrostatic interaction.

Atoms gain or lose electrons and become charged (i.e., ions), a process known as ionization.

Nonmetals (i.e., on the right side of the periodic table) tend to gain electrons and form negatively charged anions.

Metals (located on the left side of the periodic table) exhibit weaker nuclear forces on their valence electrons and lose electrons, forming positively charged cations.

Ionic bonds form between a metal cation and a nonmetal anion so that each atom obtains a complete *valence shell*.

Ionic bonds *transfer an electron* from the electropositive to the electronegative element.

Ionic bonds occur when the electronegativity difference is significant (*greater than 1.6 D*).

Ionic bonds generally occur between a metal (left two columns; groups 1 and 2) and a nonmetal (right side; generally, groups 16 and 17).

If a metal has low ionization energy and a nonmetal has a high electron affinity (discussed elsewhere), ionic bonding is more likely.

A familiar ionic compound is sodium chloride (NaCl), commonly known as table salt.

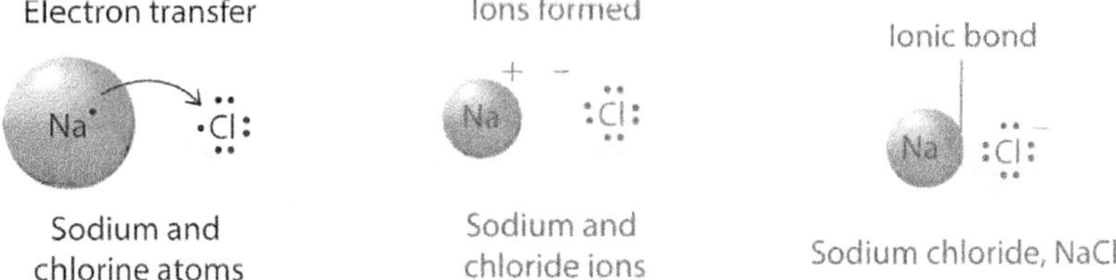

Cations and anions in an ionic compound are like a magnet's North and South poles, which attract.

Similarly, two cations (or anions) repel each other.

Crystalline structures

Therefore, the molecules arrange into rigid crystalline structures.

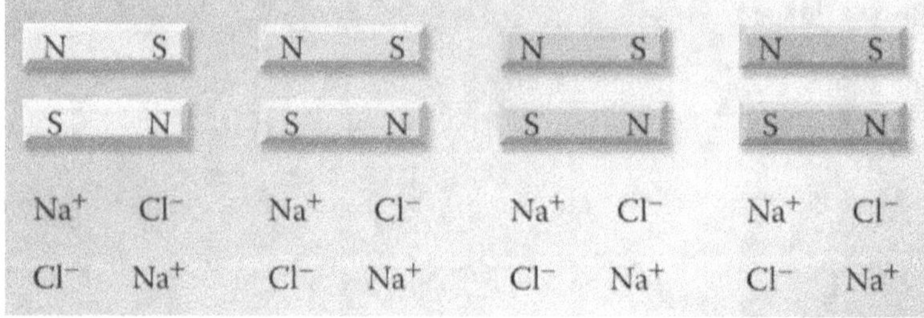

Crystalline Na⁺ and Cl⁻ with alternating positive and negative charges

Due to the strong, polarized intermolecular forces, ionic compounds generally possess high melting and boiling points. Therefore, they are often solid at room temperature.

Ionic crystals are three-dimensional arrangements of ions in an ionic compound, sometimes called a *crystal lattice* (e.g., solid crystalline structure of sodium chloride).

Sodium chloride (NaCl) has a high melting point (800 °C) and dissolves in water to give a conducting solution.

Like ionic compounds, sodium chloride is a strong electrolyte that dissociates entirely in water.

Solid ionic compounds are non-conductive; however, molten compounds (i.e., compounds liquefied by heat) or compounds dissolved in water conduct electricity.

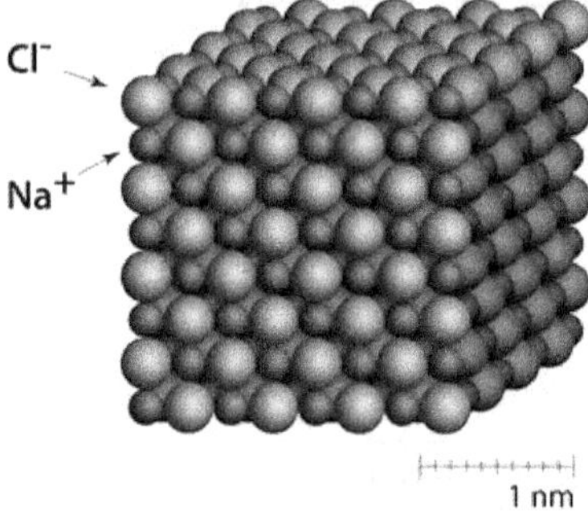

The crystalline lattice structure of table salt (NaCl)

Like all types of bonding, the valence (outermost) electrons participate in ionic bonding.

Transferring the lone $3s$ electron of a sodium atom to the half-filled $3p$ orbital of a chlorine atom generates a sodium cation and a chloride anion, forming the compound sodium chloride.

Initially, the sodium and chlorine atoms each had a net charge of 0. After the transfer of one electron took place, they had a charge of $+1$ and -1, respectively, due to the atom's imbalance between protons and electrons.

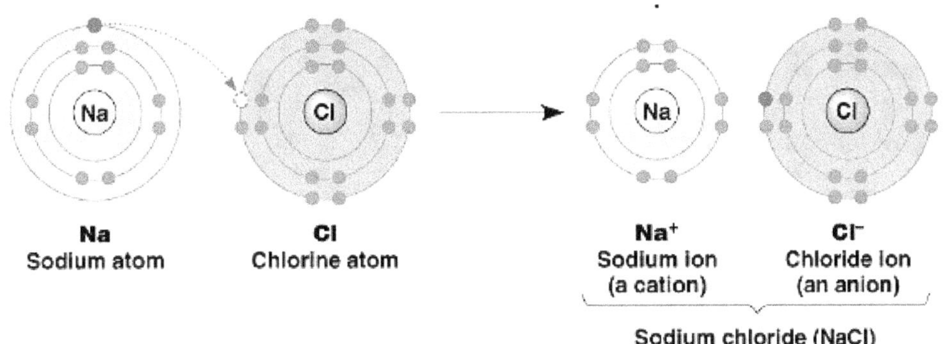

Ionic bonds form by the transfer of a valence electron

Electrostatic potential energy $\propto q_1 q_2 / r$

The strength of an ionic bond is related to the amount of *electrostatic potential energy* between opposing charges.

The *electrostatic potential energy* is:

$$U_E = k q_1 q_2 / r$$

where U_E is electrostatic potential energy, k is Coulomb's constant (9×10^9 m/F), q_1 and q_2 are the two charges, and r is the distance between them.

Electrostatic energy is negative in ionic bonds because q_1 and q_2 have opposing charges.

If q_1 and q_2 did not have opposing charges (i.e., both positive or negative), they would repel, and no ionic bond would form.

The negative sign is often omitted, and the magnitude of the electrostatic energy is used instead.

The higher the magnitude of the electrostatic potential energy, the stronger the ionic bond.

Strong ionic bonds are formed by high charge magnitudes (q values) close together (small r value).

Ions that form strong ionic bonds have a high charge density (high charge-to-size ratio).

Cation's high charge density can distort the anion's electron cloud to promote electron sharing.

Lattice energy

Lattice energy is the energy required to break an ionic bond. The energy released by the reaction correlates with the ionic bond strength, which is proportional to electrostatic energy.

Conversely, it is the energy released when oppositely charged gaseous ions join to form an ionic crystalline solid.

Electrostatic force $\propto q_1 q_2 / r^2$

The electrostatic force that holds charged particles is defined as *electrostatic energy*.

Coulomb's Law defines electrostatic force F as a function of electrostatic charges and distance.

$$F = kq_1 q_2 / r^2$$

where F is the electrostatic force, k is Coulomb's constant (9×10^9 m/F), q_1 and q_2 are the two charges, and r is the distance between the charges.

From the equation, the electrostatic force is directly proportional to the product of opposite charges that attract (negative F) or the same charges that repel (positive F).

The electrostatic force is inversely proportional to the square of the distance between them.

Therefore, larger charge magnitudes closer exhibit a greater electrostatic force.

Coulomb's Law is analogous to the *universal law of gravitation*:

$$F = Gm_1 m_2 / r^2$$

where G (gravitational constant) is analogous to k, and m (mass) to q

The major difference is that G is tiny compared to Coulomb's constant k (stationary electrically charged particles) because the gravitational force is weak compared to the stronger electrostatic force between charged particles.

There is another way to write the formula for an electrostatic force specific to ionic compounds.

The equation describes the force of attraction between the cation n^+ and the anion n^- at a distance d apart.

$$F = R(n^+ \times e) \cdot (n^- \times e) / d^2$$

where R is Coulomb's constant (usually written as k).

$n^+ \times e$ = charge of cation in coulombs = positive charge (n^+) $\times$ coulombs per electron (e)

$n^- \times e$ = charge of anion in coulombs = negative charge (n^-) $\times$ coulombs per electron (e)

Electron Pair Sharing

Covalent bond formation

Covalent bonds form when two atoms share an electron pair.

These electron pairs are shared, or bonding pairs, and covalent bonds result from the overlap of electron orbitals.

Covalent bonds typically form between nonmetal elements.

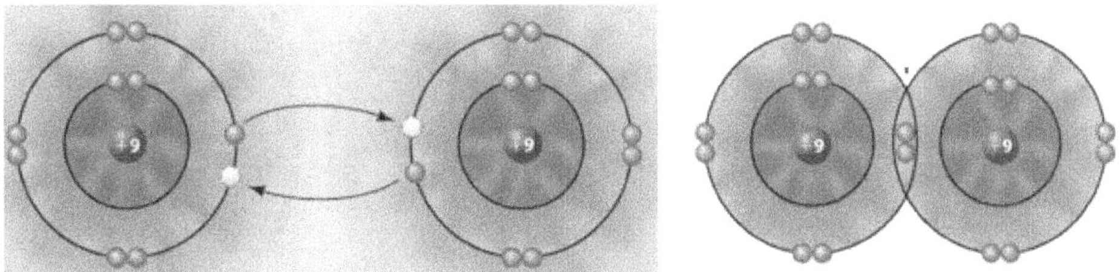

A covalent bond forms between two fluorine atoms. The left image shows the valence electron from each fluorine atom that becomes shared to form the covalent bond in the F_2 molecule

Most atoms with similar electronegativity share electrons to achieve a full valence shell of eight electrons (i.e., octet rule for $2s$ and $6p$ electrons).

Illustrations of covalent bonds following the octet rule are shown below.

Electron dot formulas

Simple atomic notation with dots designating valence electrons:

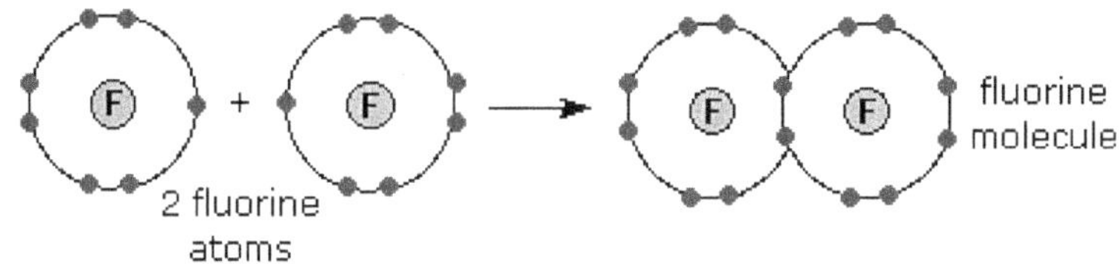

Covalent bond forms between 2 fluorine atoms that share a valence electron

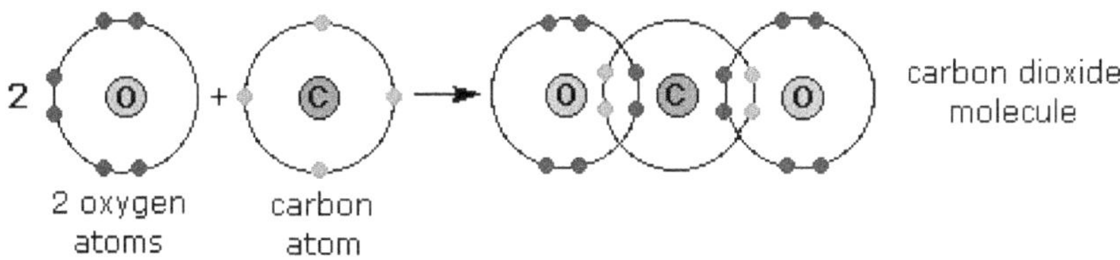

Covalent bond forms between 2 O atoms and one carbon atom

that share two pairs of valence electrons from the carbon atom to form CO_2

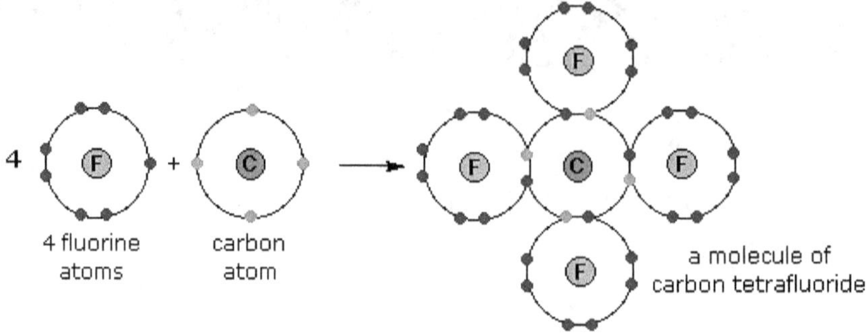

Covalent bond forms between 4 fluorine atoms and one carbon atom

Each atom contributes one valence electron to form carbon tetrafluoride (CF_4).

Hydrogen atoms are an exception to the octet rule because it is at its lowest energy when it has two $1s$ electrons in its valence shell.

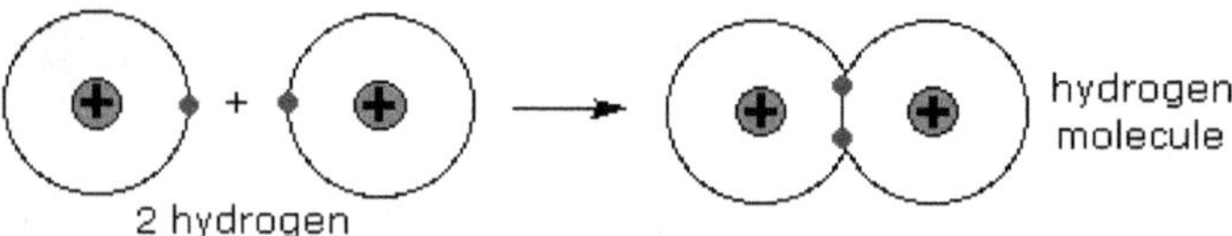

Atoms form covalent bonds to obtain the ideal number of electrons.

Non-bonding lone pairs

Non-bonding lone pairs are electrons not involved in bonding.

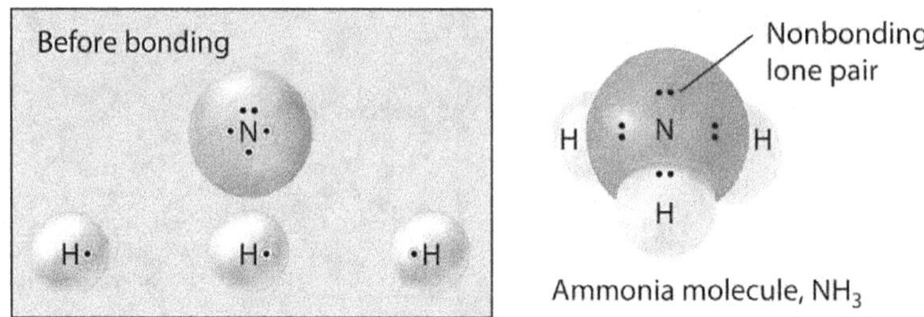

Nitrogen has 5 valence electrons. Nitrogen shares one valence electron with each of three hydrogen atoms. Each hydrogen contributes 1 valence electron. With three covalent bonds, nitrogen has a complete octet, and hydrogen satisfies a complete 1s subshell.

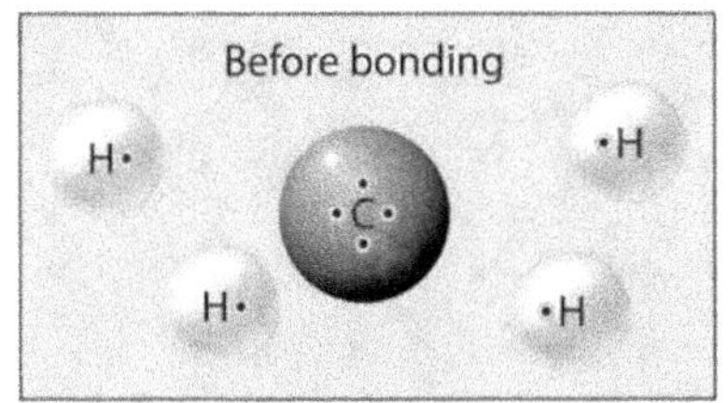

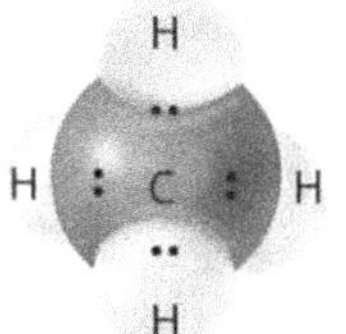

Methane molecule, CH_4

Carbon has 4 valence electrons and shares one electron with each of the four H atoms. Each H contributes 1 valence electron. With four covalent bonds, carbon has a complete octet, and H satisfies a complete 1s subshell

Multiple bonds

Covalent bonds in double- and triple-bonded nonmetal atoms involve more than one pair of electrons.

In double bonds, *two pairs* of electrons (four electrons) are shared.

In triple bonds, *three pairs* of electrons (six electrons) are shared.

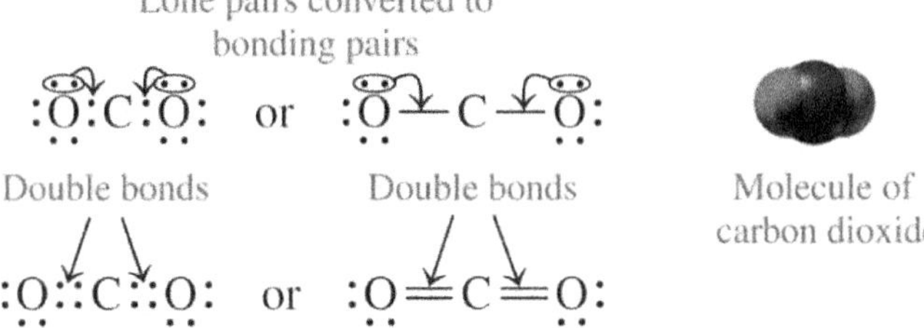

Carbon dioxide forms from two double bonds on the central carbon. Each double bond results from the oxygen and the carbon contributing a pair of electrons

$$·N:N· \longrightarrow :N::N: \quad :N≡N:$$

Three shared pairs   Triple bond N_2 molecule

Coordinate covalent bonds

Coordinate covalent bonds result from one atom donating an electron pair to another atom; both electrons come from one donor atom.

Ozone, O_3, is a molecule with a coordinate covalent bond.

$$:O::O: + O: \longrightarrow :O::O:O:$$

nonbonding electron pairs   coordinate covalent bond: O=O–O

Notes for active learning

Sigma and *Pi* Bonds

Single bonds

The electron density of a chemical bond can be divided into regions.

Sigma (σ) *bonds* form when one orbital from each atom overlaps and creates a new orbital.

Typically, a *sigma* bond is a single bond and is the first bond involved in double and triple bonds.

Sigma bonds lie along the imaginary line connecting nuclei, known as the *internuclear axis*.

In σ bonds, the electron density is between the two atoms' nuclei and is symmetrical about the internuclear axis.

Atoms on each side of a σ single bond are free to rotate with the bond as their axis.

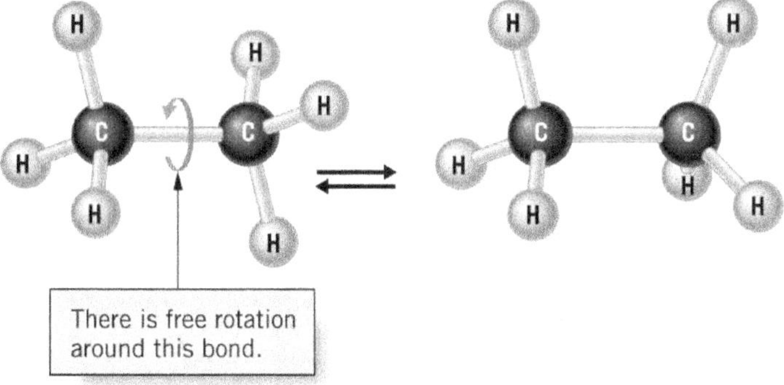

S and *p* orbitals

In the image below, the σ bonds of *s*–*s* and *p*–*p* orbitals are shown.

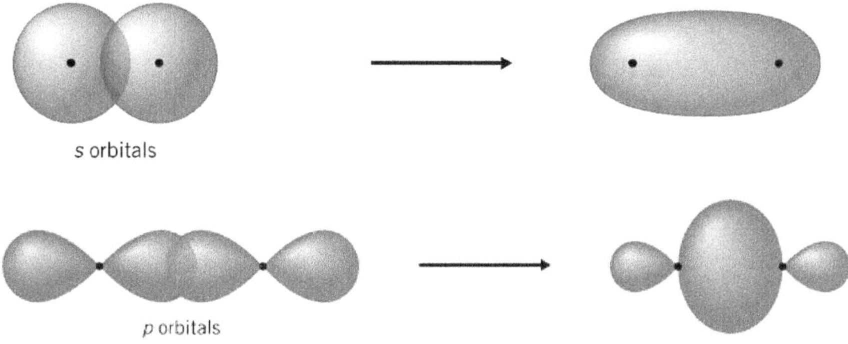

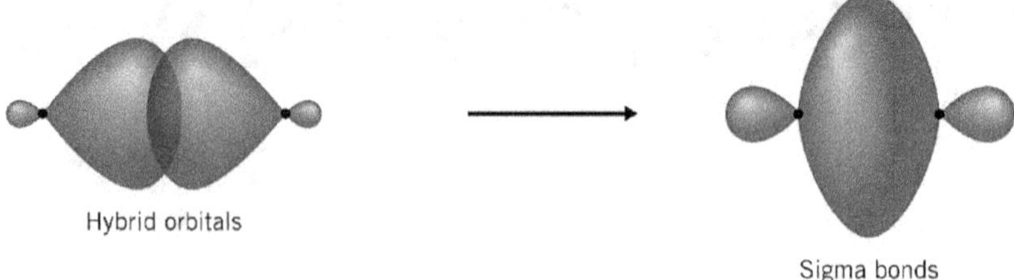

Double bonds

Pi (π) bonds form when unhybridized *p* orbitals on the top and bottom of the internuclear axis overlap.

A single π bond consists of regions above and below the intermolecular axis.

There is no electron density along the π bond intermolecular axis.

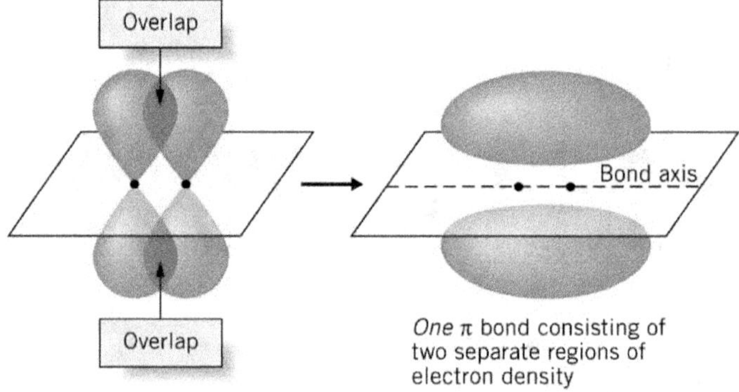

When electron density above and below the σ bonds overlap, a multiple bond (i.e., double or triple) forms. Thus, π bonds form after a σ bond has connected two atoms.

Double bond: one σ bond and one π bond.

Triple bond: one σ bond and two π bonds.

Nitrogen (N_2) gas is connected by a triple bond ($N\equiv N$). The first is a σ bond, and the remaining two are π bonds.

π bonds introduce rigidity to molecules; π bonds do not allow for the rotation of atoms.

Molecules that involve π bonds require more energy to break.

Hybridization

Hybrid orbitals

Hybrid orbitals are produced by the hybridization (mixing) of existing electron orbitals.

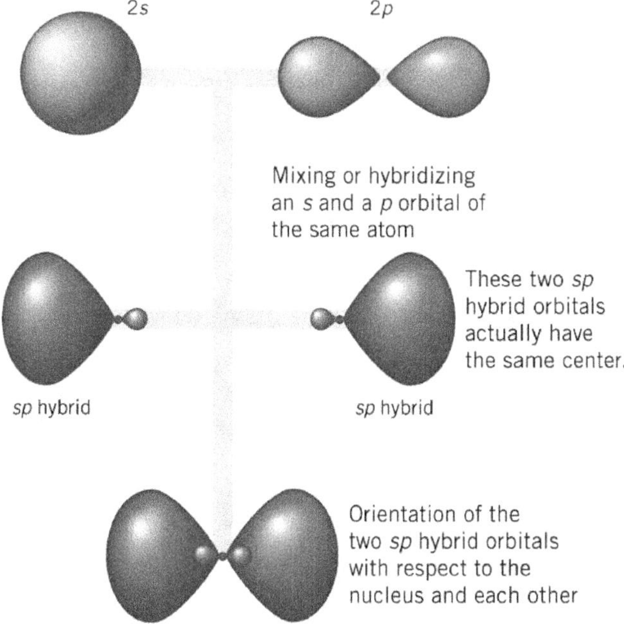

Hybridization and molecular geometry

Hybridizations of the carbon atom are as follows.

- sp^3: a hybrid between one s with three p orbitals.

  Tetrahedral geometry.

  Contains single bonds only.

- sp^2: a hybrid between one s with two p orbitals.

  Trigonal planar geometry.

  Contains a double bond.

- sp: a hybrid between one s with one p orbital.

  Linear geometry.

  Contains a triple bond.

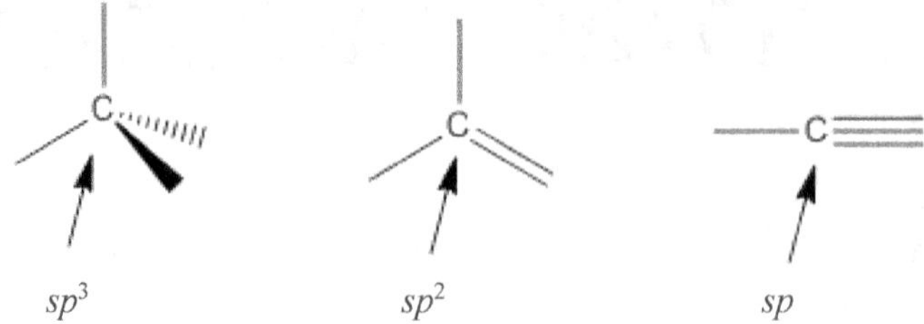

Hybridization of the carbon atom for 4 single bonds, 1 double or 1 triple bond

Valence shell electron-pair repulsion (VSEPR) theory predicts molecular shapes

Valence Shell Electron Pair Repulsion (VSEPR) theory states that electron pairs surrounding an atom repel each other and have the greatest bond angles separating the bonds and lone pairs around a central atom.

In VSEPR, *electron domains* are electron pairs in the molecule.

Molecules have *bonding* and *non-bonding domains*.

When atoms bond to create molecules, the atoms arrange as far apart as possible to minimize the same-charge repulsion between electrons.

VSEPR theory predicts the *molecular geometry* based on the *number of electron pairs* surrounding the central atom.

Electron pair domain geometry (EDG) indicates bonding and non-bonding electron pairs around the central atom.

Molecular shape geometry (MG) indicates the arrangement of atoms around the central atom, considering electron repulsion.

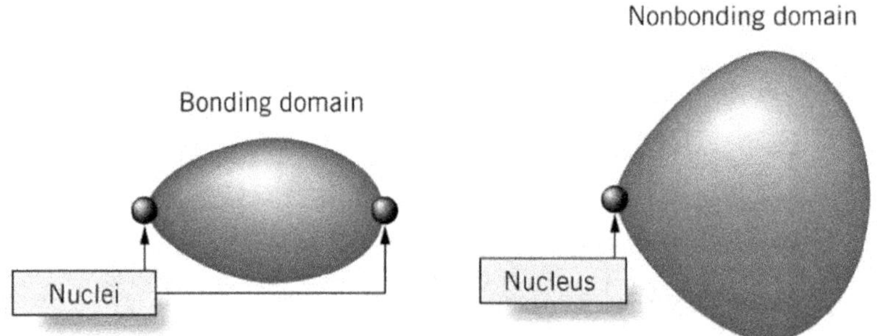

Bonding domains are directed between two adjacent nuclei, while the antibonding domains are directed away from adjacent nuclei.

Shape and bond angle

Specific configurations (i.e., specific shapes) maximize the distance between the bonding electron pairs and substituents attached *via* the bond.

Molecules with *two* attached atoms (or substituents) are *linear* (a straight line, with one atom on each side of the central atom).

Bond angle = 180°

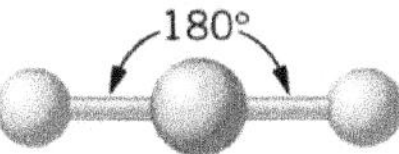

Molecules with *three* attached atoms are *trigonal planar*.

Bond angle = 120°

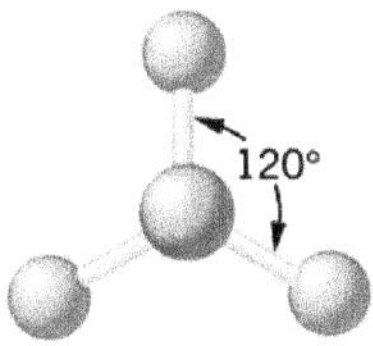

Molecules with *four* attached atoms form a *tetrahedron*.

Bond angle = 109.5°

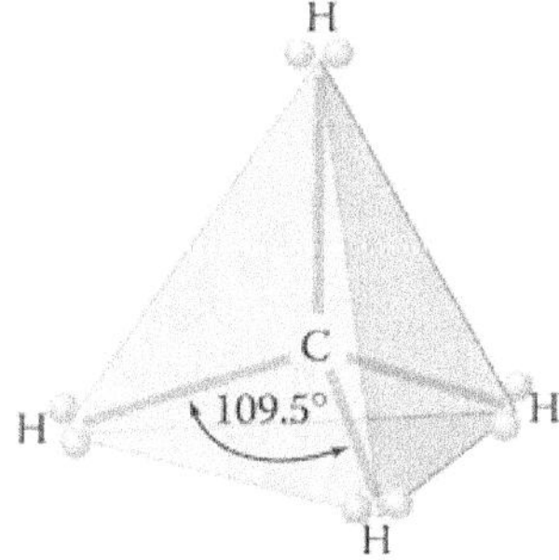

Molecules with *five* attached atoms are *trigonal bipyramidal*.

Bond angles = 120° and 90°

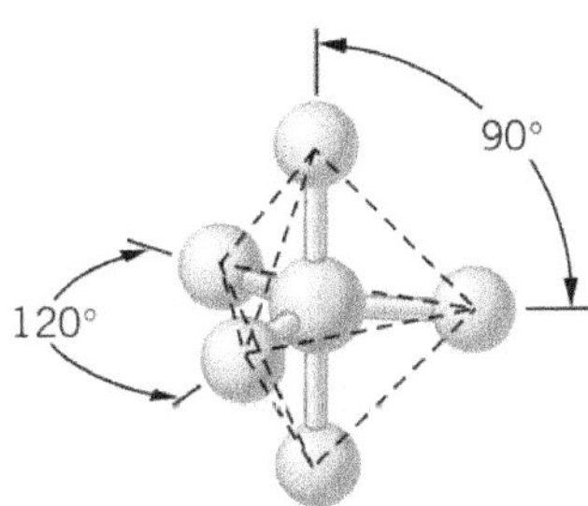

Molecules with *six* attached atoms are an *octahedron*.

Bond angle = 90°

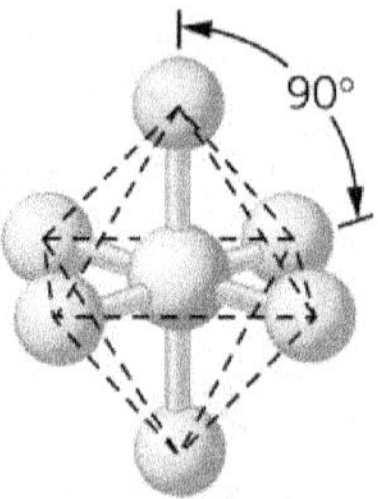

Molecular shapes above apply to molecules containing only bonding domains (i.e., central atoms do not have non-bonding domains or lone pairs of electrons).

Non-bonding domains

When the central atom has non-bonding domains, the lone pair occupies the space where a bond usually forms.

These non-bonding electron domains are relatively dispersed in space and exert electrostatic repulsions that cause bonding electrons to bend away.

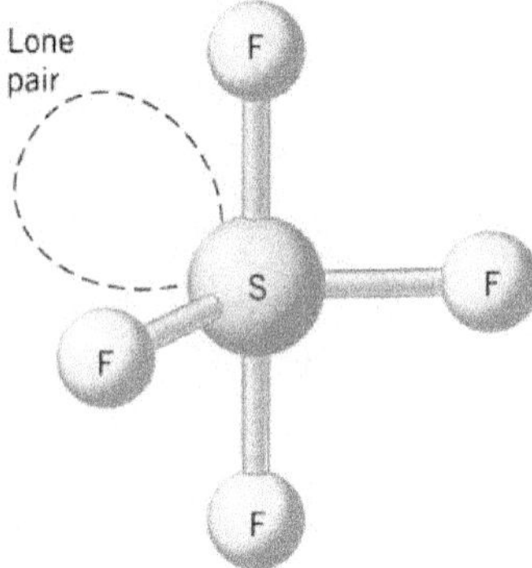

Lone pairs must be closer to the atom whose orbital they occupy than the bonding pairs shared along the two bonding atoms' internuclear axis.

Thus, lone pairs are not restricted to the internucleus axis and occupy more available space within the orbital (i.e., more electrostatic repulsion).

An example is SF_4, which has one lone pair:

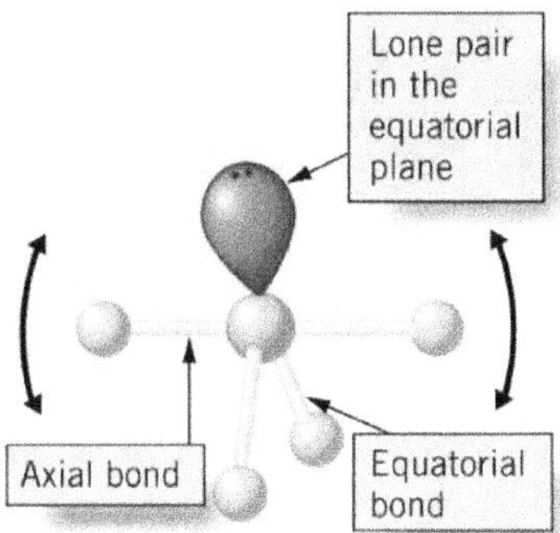

Electrons in sigma bonds occupy a restricted space, while nonbonding lone pairs are more dispersed and therefore contribute electrostatic repulsion to the bonding electrons

Hybridization, bond angles, and molecular geometry rules

Hybrid ization	Electron geometry	Electron groups	Bonding groups	Lone pairs	Approx. bond angles	Molecular geometry
sp	Linear	2	2	0	180°	Linear
sp²	Trigonal planar	3	3	0	120°	Trigonal planar
		3	2	1	< 120°	Bent
sp³	Tetrahedral	4	4	0	109.5°	Tetrahedral
		4	3	1	< 109.5°	Trigonal pyramidal
		4	2	2	<< 109.5°	Bent
sp³d	Trigonal bipyramidal	5	5	0	120° (equatorial) 90° (axial)	Trigonal bipyramidal
		5	4	1	< 120° (equatorial) < 90° (axial)	Seesaw
		5	3	2	< 90°	T-shaped
		5	2	3	180°	Linear
sp³d²	Octahedral	6	6	0	90°	Octahedral
		6	5	1	< 90°	Square pyramidal
		6	4	2	90°	Square planar

Determining the molecule's geometric shape

1. Count the domains (bonding and non-bonding) around the central atom.

    The number of domains determines the molecule's general shape (above).

    For every non-bonding domain, remove the surrounding atom.

    For example, a four-domain molecule has a tetrahedral shape.

    If one of the domains is non-bonding, one of the surrounding atoms is removed, leaving a central atom with three surrounding atoms; the molecule has a trigonal pyramid shape.

    If there are two non-bonding domains, remove two surrounding atoms; the molecule is a bent shape.

2. For molecules with five domains, start by removing atoms above and below the pyramid before the 3 atoms surrounding it.

    A molecule is more stable when its atoms adopt a configuration that minimizes repulsion.

    Angles between the top or bottom atoms and the three side atoms are 90°.

    Angles between the side atoms are 120°.

    Molecules shed the more repulsive atoms first (i.e., those with smaller angles).

    The same approach applies to molecules with six domains.

    Domains above and below are removed before the central atom domains are considered.

Non-bonding domains affect the angle of bonds in the molecule due to occupying more significant space (increased electrostatic repulsion) than bonded electrons.

The following diagram summarizes the shapes of atoms based on the number of bonding and non-bonding electron pairs (also called nonbonded electron pairs):

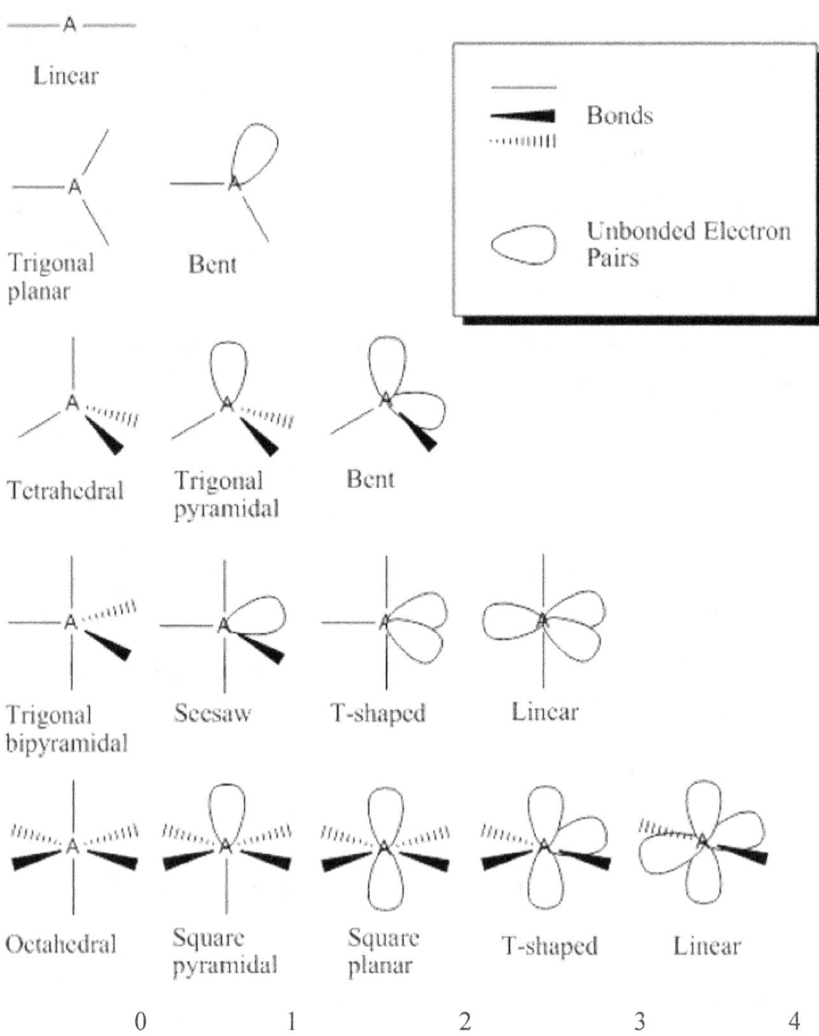

Numbers along the bottom indicate the # of nonbonded electron pairs

Oxygen of water (H_2O, shown below) has two bonding and two non-bonding domains. The water molecule experiences electron repulsion from the two lone pairs on its bonds.

Water molecules adopt a bent shape (see prior chart) to minimize the van der Waals repulsion of the negatively charged electrons.

Water's bent shape comes from a tetrahedral configuration, with two vertices removed. Bonds in a tetrahedron are 109.5°, which means bonds in a bent molecule should be 109.5°.

However, because of the repulsion between lone electron pairs and the bonding electrons, the reported bond angle is slightly decreased to approximately 105°.

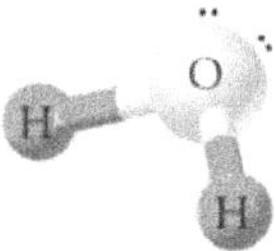

A water molecule (H₂O) is bent due to the two lone pairs of e⁻ on oxygen

VSEPR geometric shapes classify double bonds as a single domain.

Formaldehyde (CH_2O) has 1 double and 2 single bonds, resulting in three bonding domains.

The molecular geometry of formaldehyde is trigonal planar (i.e., flat shape with ~120 bonds).

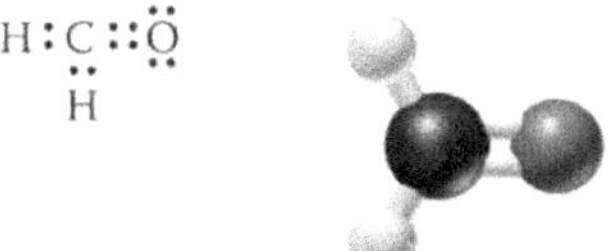

Trigonal plane geometry for formaldehyde (CH₂O) with 1 double (π) and 2 single (σ) bonds

Two major theories explain how molecules bond: valence bond theory and molecular orbital theory. Both theories are used to predict the structure of a molecule.

Notes for active learning

Valence Bond (VB) Theory

Atomic orbital (AO) theory

Valence bond theory (VB theory) is primarily based on valence electrons and states that each atom has an orbital, and *orbitals overlap to form bonds*.

The extent of overlap of atomic orbitals (AO) is related to bond strength; more overlap results in a stronger bond.

VB theory explains the bonding of F_2. F_2 bonds form from atomic valence orbitals overlapping (e.g., $2p$ orbitals), one from each fluorine.

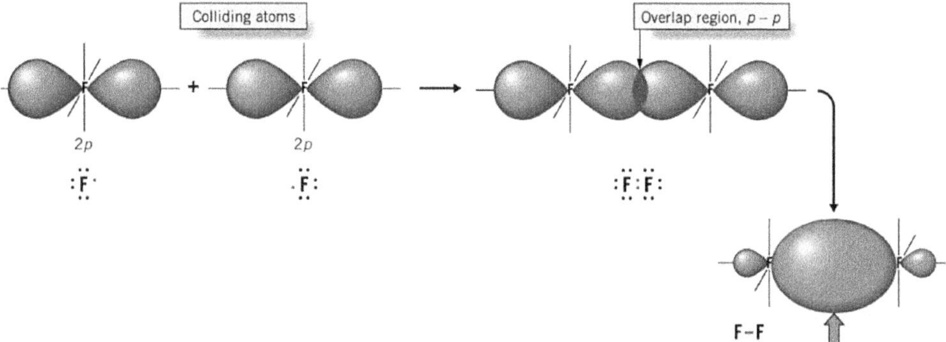

The overlap of p–p orbitals forms the F₂ bond

Orbital overlap is not limited to identical orbitals.

Bonding in HF molecules uses the overlap between the $1s$ orbital of H and the $2p$ orbital of F.

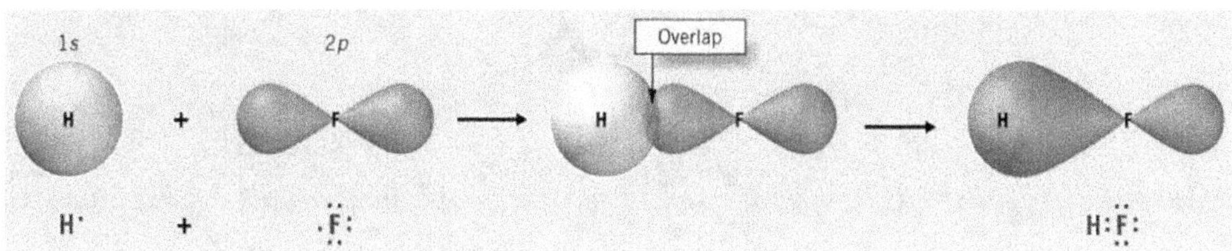

The bonding of H_2S can be described according to VB theory. Sulfur has one filled orbital (which contains two electrons) and two partially filled orbitals (each containing one electron).

Partially filled p orbitals overlap with hydrogen's s orbital.

The predicted 90° bond angle is close to the measured value of 92°.

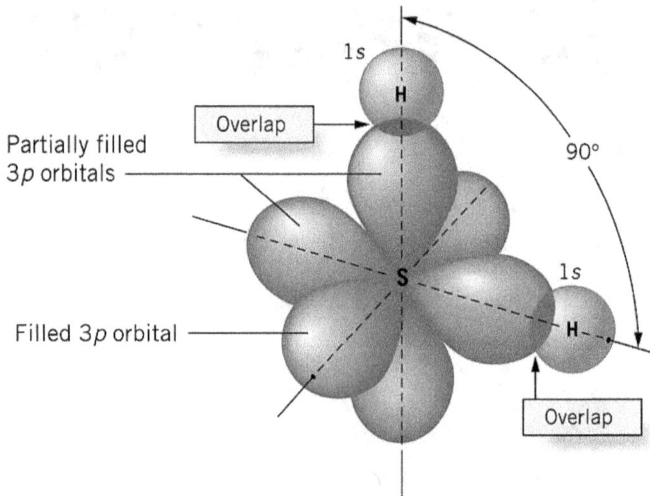

VB theory describes the bonding of many molecules, but it cannot explain the existing molecular geometries.

Methane (CH_4) has a tetrahedral shape, seen in the image below.

Carbon's electron configuration ($1s^2 2s^2 2p^2$) indicates that its electrons are paired, except for two electrons that occupy p orbitals.

VB theory predicts that carbon can only bond with two hydrogen molecules, resulting in a CH_2 molecule, and that the bond angle is approximately 90° (as seen in H_2S above).

Observations of methane reveal four equal bonds in the shape of a tetrahedron with bond angles of 109.5°.

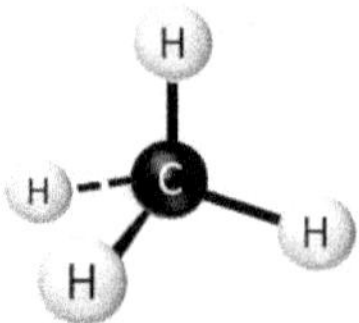

Methane (CH_4) adopts a tetrahedral shape to maximize repulsion from the 4 H substituents.

Hybridization mixes atomic orbitals

The *hybridization* concept enhanced VB theory to reconcile theoretical and experimental value differences.

Hybridization is the mixing of *atomic orbitals* (AO) to form bonds with unexpected angles.

Atomic orbitals have the highest probabilities of finding electrons, and *hybridization* is a rearrangement of orbitals.

Hybrid orbitals exhibit new shapes, directional properties, and are composed of combined constituent orbitals (shown below).

Hybrid orbitals combine the symbols of the initial orbitals.

The sum of exponents in hybrid orbital notation equals the number of atomic orbitals used.

- One s and one p orbital form two sp hybrid orbitals

- One s and two p orbitals form three sp^2 hybrid orbitals

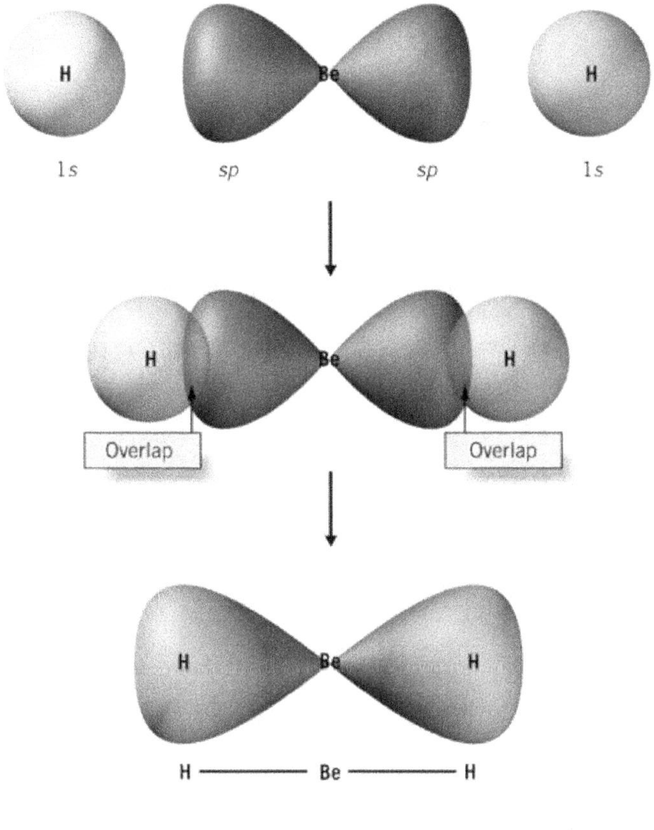

Bonding of BeH$_2$ Be: $1s^2 2s^2$ H: $1s^1$

Beryllium (Be) has two electrons in the $2s$ orbital. When bonding with two hydrogens, it splits the electrons into two orbitals. Be atom uses the next available orbital ($2p$) in combination with the $2s$ orbital, resulting in two sp orbitals.

Each of those sp orbitals overlaps with hydrogen's $1s$ orbital, creating a covalent bond.

VSEPR theory predicts that this molecule has a linear shape, which is confirmed by experimental observations.

Most atoms form up to four bonds (i.e., octet rule); four orbitals are needed (1 s and 3 p).

Bonding and non-bonding pairs

For molecules with more than four bonds/lone pairs around them, *d* orbitals are included in the hybridization to accommodate the extra bonds, known as *expanded octet hybridization.*

Bond angle experiments suggest that NH_3 and H_2O use *sp*^{3} hybrid orbitals in bonding (a total of four orbitals) even though each has two bonds.

This evidence for *sp*^{3} hybrid orbitals suggests that hybrid orbitals are not exclusively used for bonding; they also accommodate non-bonding electron pairs.

Bonding and non-bonding pairs contribute to the molecule's geometry.

For example, the orbital diagram of hybridized nitrogen in NH_3:

Hybridization accommodates 3 bonding and 1 nonbonding electron pairs

Diagram showing the hybridization of oxygen in H_2O:

Hybridization accommodates 2 bonding and 2 nonbonding electron pairs

Molecular Orbital (MO) Theory

Wave interference

Molecular orbital theory (MO theory), the second dominant theory, views molecules as a collection of positively charged nuclei having a set of molecular orbitals filled with electrons (like filling atomic orbitals with electrons).

It does not focus on how individual atoms join to form molecules.

The MO theory is more difficult to visualize than VB theory.

MO theory is advantageous because it accurately predicts a molecule's *magnetic properties* and other characteristics.

Energies of molecular orbitals (MO) are determined by combining *atomic orbitals'* (AO) *electron waves*.

For H_2 (below), two $1s$ wave functions, one from each atom, combine to make two MO wave functions.

Two MOs produce *constructive interference* of the waves.

Bonding and antibonding molecular orbitals

For H_2, the bonding MO's energy is lower than the atomic orbitals' energy (AO), and the molecule is collectively more stable than the individual atoms (atoms forming the molecule).

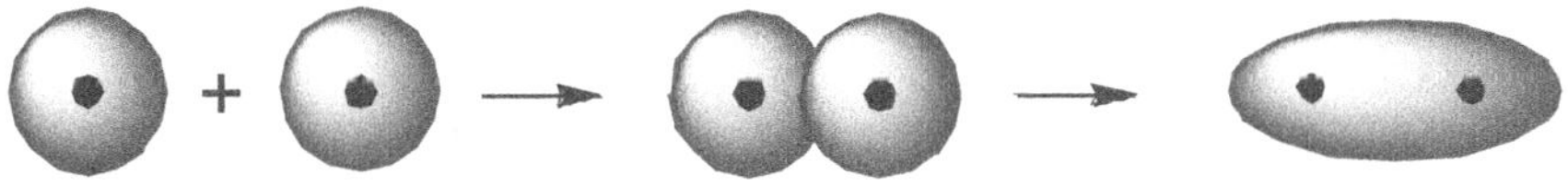

As there are σ orbitals, there are corresponding σ^* antibonding MOs.

The possible combination of two $1s$ orbitals involves destructive interference of the $1s$ waves.

In the example below, the antibonding MOs' energy is higher than the energy of the parent atomic orbitals. The molecule *does not form*, and the atoms remain separate.

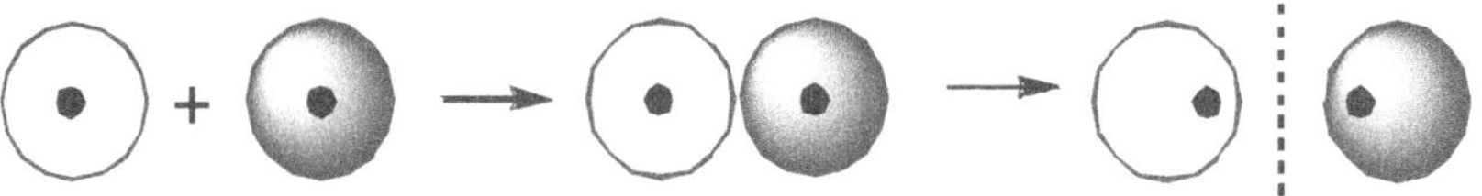

Antibonding MOs are high in energy and do not contribute to bond formation.

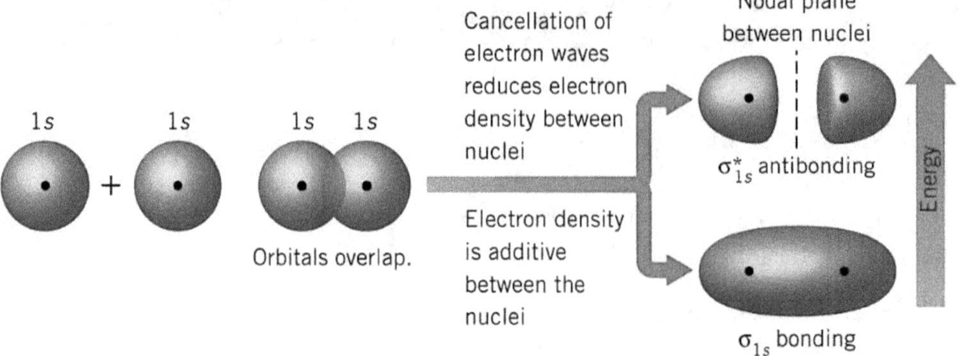

Bonding and antibonding MOs with the relative energy of each

Bonding MOs have electron density between nuclei; the electrons tend to stabilize the molecule.

Antibonding MOs cancel electron waves, reducing the nuclei' electron density.

Electrons in antibonding MOs tend to destabilize the molecule.

Molecular orbital theory diagrams

MO energy diagrams (shown below) represent the interactions between atomic orbitals.

MO energy diagrams display the orbitals arranged vertically, from highest to lowest energy.

Atomic orbitals (AO) for the atoms are on the left and right sides of the diagram.

Molecular orbitals (MO) are in the central column of the diagram.

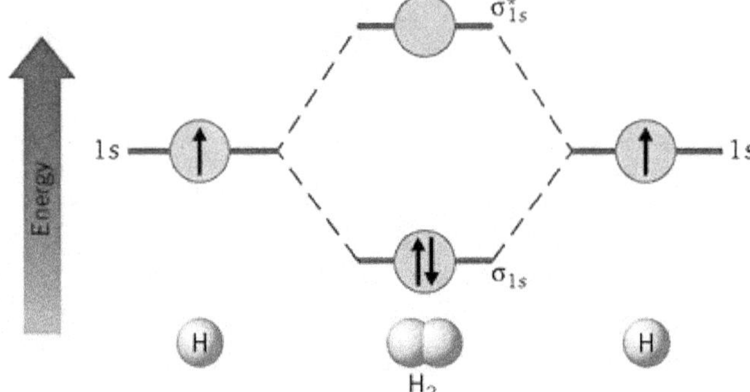

MO energy diagram for H_2: atomic orbitals are designated 1s and have single arrows up, while the lower energy bonding molecular orbital has 2 electrons of opposite spin.

The high-energy antibonding MO is *vacant*. The molecule forms from separate atoms because the bonded molecule is *more stable* than the separate atoms (relative stability, as indicated by the energy on the y-axis).

Three rules for filling MO energy diagrams:

1. Electrons fill the lowest-energy orbital (i.e., Aufbau's principle).

2. No more than two electrons (with opposite spins) can occupy any orbital (i.e., Pauli exclusion principle).

3. Electrons with unpaired spins spread over the same energy orbitals (i.e., Hund's rule).

Bond order

Bond order indicates the number of electron pairs shared between atoms (i.e., the number of bonds between two atoms).

Bond order of one corresponds to a single bond.

For example, the H_2 bond order and the C−H bond order are each one, while the N≡N bond order in diatomic nitrogen is three.

Bond order is calculated by:

$$\text{Bond order} = \frac{(\text{number of bonding } e^-) - (\text{number of antibonding } e^-)}{2 \text{ electrons / bond}}$$

Valence bonding and molecular orbital theory limitations

Neither *valence bonding* (VB) nor *molecular orbital theory* (MO) is entirely correct because neither theory explains all aspects of bonding; each theory has its strengths and weaknesses.

VB theory is based on Lewis structures (discussed later) and the related geometric shapes.

MO theory correctly predicts the unpaired electrons in O_2, while the Lewis structures do not.

Molecular orbital theory is complicated because even simple molecules have complex energy level diagrams, and molecules with three or more atoms require extensive calculations.

Valence bond theory utilizes three-dimensional structures based on electron domains, eliminating the need for extensive calculations.

Simple hybrid orbitals are invoked when experimental evidence indicates need, and the integer bond orders are often accurate.

Structural formulas for heteroatom molecules

Lewis structures (explained later) are used to visually represent atoms and their bonds to other atoms in molecules.

Lewis structures for elements in the same column (or group) are similar.

For example, sulfur can be substituted for oxygen in the Lewis structures of oxygen.

Below are structural formulas for some important molecules.

Hydrogen Lewis structures

hydrogen proton: H^+

hydride ion: H^-

Boron – Group 13 Lewis structures

Since group 13 elements have three valence electrons, they often have six total electrons in a molecule (an exception to the octet rule).

Borane:

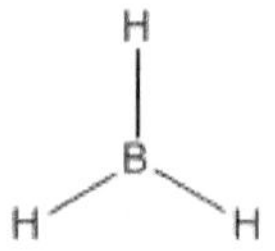

Borohydride ion:

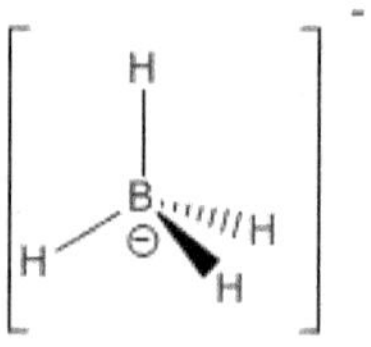

Borohydride is expressed as BH_4^- with a formal charge on boron (B), shown with brackets and a negative charge on the top right side.

Carbon – Group 14 Lewis structures

Methane:

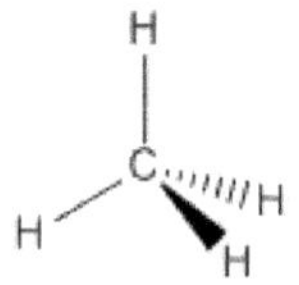

Carbocation:

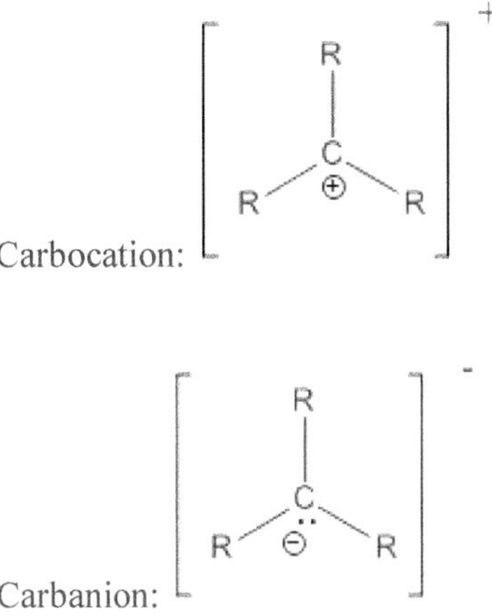

Carbanion:

Note: R in the figures is carbon or hydrogen.

Nitrogen – Group 15 Lewis structures

Some group 15 elements have more than eight electrons after bonding, an exception to the octet rule (e.g., PCl_5).

Presence of *d* orbitals permits the atom to accommodate these additional bonds (more than 4).

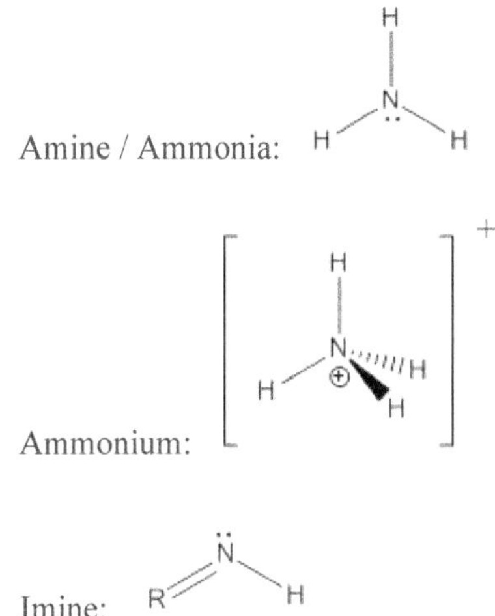

Amine / Ammonia:

Ammonium:

Imine:

Oxygen – Group 16 Lewis structures

Some group 16 elements have more than eight electrons after bonding, an exception to the octet rule (e.g., SF_6).

Molecular oxygen:

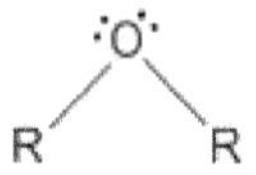

Water, alcohol, and ethers:

Ozone:

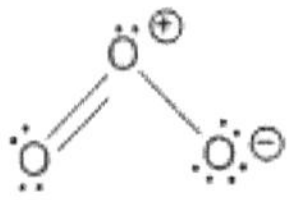

Halogen – Group 17 Lewis structures

Hydrogen fluoride:

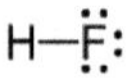

Chloromethane:

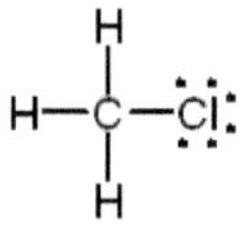

Bromide ion:

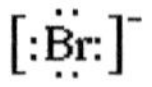

Lewis Electron Dot Formulas

Lewis structures

Lewis dot formulas, or *Lewis structures*, are notations used to describe bonds and electrons in a molecule.

For Lewis structures, each *dot represents one electron*, and each line represents one bond (two electrons); two dots represent a lone pair.

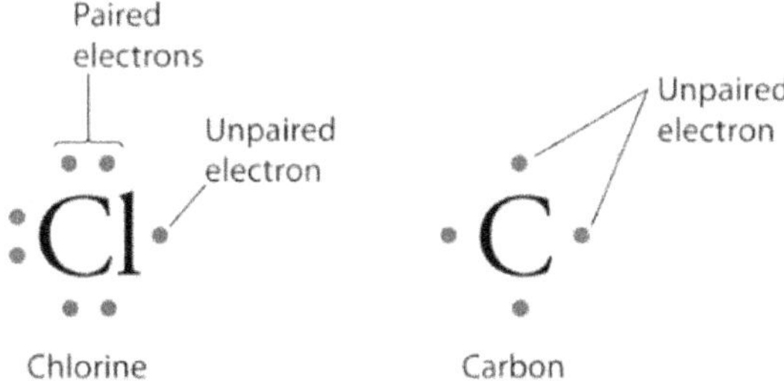

Lone pairs are represented as two dots, while an unpaired electron has no partner.

$$Cl:P:Cl \quad \text{or} \quad Cl-P-Cl$$

Lone pairs are usually omitted from the molecule's final structure:

$$:\ddot{F}-\ddot{F}: \qquad F-F$$

Recalling the *octet rule*: most atoms need a total of eight valence electrons to achieve stable electron configurations.

Electrons in a bond are shared and can satisfy the bonded atoms' octet.

This electron configuration principle includes coordinate covalent bonds, where both bonding electrons are contributed by one atom.

However, the electron pair counts towards the other atom's octet rule.

Exceptions for the octet rule include atoms with other than 8 e⁻:

> boron column (they form three bonds and have a sextet),

> large elements with periods of 3 or higher, where these elements have d orbitals (e.g., the dectet P in PO_4^{3-} and the duodectet S in SO_4^{2-}),

> radicals (compounds with an odd number of total electrons possess a single, unpaired electron).

General rules for Lewis structures

- From the octet rule, most atoms have up to eight electrons.

- For atoms with up to four valence electrons, place them evenly around the molecule.

- If the atom has five or more valence electrons, divide them into four sections with a maximum of two electrons on each side. This is not required but makes it easier to determine if an atom has achieved stability (e.g., eight electrons).

When sketching Lewis structures, use distinct symbols to represent electron sources.

For example, consider the Lewis structure of PCl_3 below; it is unclear which electrons on phosphorus (P) originated from phosphorus or chlorine (Cl).

P has five valence electrons; deduce that the remaining three come from Cl.

However, it can be confusing when the bonds get more complicated (e.g., SO_3).

Use clear designations to differentiate electrons by source.

In $BeCl_2$ below, "x" represents electrons from Be, and dots electrons from Cl.

Place square brackets around complete Lewis structures and indicate the charge on the top of the right bracket.

Drawing electron dot formulas

1) Determine the *arrangement* of the atoms.

2) Determine the total number of *valence electrons*.

3) Attach a bonded atom to the *central atom with a pair of electrons*.

4) Place the remaining electrons using *single or multiple bonds to complete the octet*.

Electron-dot formulae using valence electrons

Molecule or Polyatomic Ion	Total Valence Electrons	Form Single Bonds to Attach Atoms (electrons used)	Electrons Remaining	Completed Octets (or H:)
Cl_2	$2(7) = 14$	$Cl—Cl\ (2\,e^-)$	$14 - 2 = 12$	$:\ddot{C}l—\ddot{C}l:$
HCl	$1 + 7 = 8$	$H—Cl\ (2\,e^-)$	$8 - 2 = 6$	$H—\ddot{C}l:$
H_2O	$2(1) + 6 = 8$	$H—O—H\ (4\,e^-)$	$8 - 4 = 4$	$H—\ddot{O}—H$
PCl_3	$5 + 3(7) = 26$	$Cl—P(—Cl)—Cl\ (6\,e^-)$	$26 - 6 = 20$	$:\ddot{C}l—P(—\ddot{C}l:)—\ddot{C}l:$
ClO_3^-	$7 + 3(6) + 1 = 26$	$\left[O—Cl(—O)—O\right]^-\ (6\,e^-)$	$26 - 6 = 20$	$\left[:\ddot{O}—\ddot{C}l(—\ddot{O}:)—\ddot{O}:\right]^-$
NO_2^-	$5 + 2(6) + 1 = 18$	$\left[O—N—O\right]^-\ (4\,e^-)$	$18 - 4 = 14$	$\left[:\ddot{O}—\ddot{N}=\ddot{O}:\right]^- \longleftrightarrow \left[:\ddot{O}=\ddot{N}—\ddot{O}:\right]^-$

Lewis structure rules for common elements

Carbon: four bonds, zero lone pairs. (e.g., CH_4, CO_2)

Oxygen:

O: two bonds total, 2 lone pairs (e.g., H_2O, O_2)

O^{1-}: one bond, 3 lone pairs, a formal charge of -1

O^{1+}: three bonds, 1 lone pair, a formal charge of $+1$

Nitrogen:

N: three bonds total, one lone pair (e.g., amines, ammonia NH_3)

N^+: four bonds, zero lone pairs, a formal charge of $+1$ (e.g., ammonium NH_4^+)

Halogens: one bond, three lone pairs (e.g., CCl_4)

Hydrogen: one bond, zero lone pairs (octet rule exception)

Carbocation: C^+ has three bonds, no lone pairs

Carbanion: C^- has three bonds, one lone pair

Boron: three bonds, zero lone pairs (an exception to the octet rule) (e.g., BH_3)

Formal Charge

Formal charge for resonance structures

Formal charge is assigned to each atom in a Lewis structure based on equal sharing of bonded electron pairs.

Formal charges are used in drawing resonance structures (multiple Lewis structures that collectively describe a single molecule) of covalently bonded molecules.

This technique describes and compares resonance structures, like how oxidation numbers balance chemical equations in oxidation-reduction reactions.

Formal charge is calculated by:

Formal charge = valence e^- in a neutral atom – (unshared valence e^- + ½ shared e^-)

where e^- represents electrons.

Every atom in a molecule has a formal charge.

In a neutral molecule, the sum of atomic charges is zero. That does not mean the atoms are not charged; they cancel each other out to create a neutral molecule.

Charge is *not evenly distributed* in charged molecules, and formal charges can help determine where the charge rests.

Calculating formal charge

Formal charge = valence e^- in a neutral atom – (unshared valence e^- + ½ shared e^-)

For Lewis structures:

Formal charge = [# of valence e^-] – [dots around atom + lines connected to atom]

Number of valence electrons is generally equal to the atom's group number on the periodic table (e.g., N has five valence electrons, O has six valence electrons, and F has seven valence electrons).

Dots around the atom represent electrons held entirely by the atom.

The lines connecting the atoms represent bonding electron pairs; each atom receives one of the two electrons.

Label the atom with the formal charge for formal charges other than zero.

Examples of atoms with formal charges:

Oxygen with a single bond: −1

Oxygen with no bonds but with an octet: −2

Carbon with three bonds:

+1 as a carbocation

−1 as a carbanion

Nitrogen with four bonds: +1

Nitrogen with three bonds: −1

Halogen with no bonds but has an octet: −1

Boron with four bonds: −1. (e.g., $^{-}BH_4$)

Carbon:

$$=C= \qquad -C\equiv \qquad \diagup_{\diagdown}C= \qquad -\overset{|}{\underset{|}{C}}- \qquad \text{4 covalent bonds: Formal charge} = 0$$

Nitrogen:

$$=\overset{\oplus}{N}= \qquad -\overset{\oplus}{N}\equiv \qquad \diagup_{\diagdown}\overset{\oplus}{N}= \qquad -\overset{\oplus}{\underset{|}{\overset{|}{N}}}- \qquad \text{4 covalent bonds: Formal charge} = +1$$

$$-\ddot{N}= \qquad :N\equiv \qquad -\overset{\cdot\cdot}{\underset{|}{N}}- \qquad \text{3 covalent bonds, 1 lone pair: Formal charge} = 0$$

$$-\overset{\cdot\cdot}{\underset{\cdot\cdot}{N}}\overset{\ominus}{} \qquad :\overset{\ominus}{N}= \qquad \text{2 covalent bonds, 2 lone pairs: Formal charge} = -1$$

Oxygen:

$$-\overset{\oplus}{\underset{\cdot\cdot}{O}}= \qquad :\overset{\oplus}{O}\equiv \qquad -\overset{\oplus}{\underset{|}{\overset{\cdot\cdot}{O}}}- \qquad \text{3 covalent bonds, 1 lone pair: Formal charge} = +1$$

$$-\overset{\cdot\cdot}{\underset{\cdot\cdot}{O}}- \qquad \overset{\cdot\cdot}{\underset{\cdot\cdot}{O}}= \qquad \text{2 covalent bonds, 2 lone pairs: Formal charge} = 0$$

$$\overset{\ominus}{}:\overset{\cdot\cdot}{\underset{\cdot\cdot}{O}}- \qquad \text{1 covalent bond, 3 lone pairs: Formal charge} = -1$$

Valence number is the number of outer shell (valence) electrons an atom needs to gain or lose to achieve a complete octet.

In covalent compounds, the number of bonds characteristically formed by an atom equals the valence number.

Atom	H	C	N	O	F	Cl	Br	I
Valence	1	4	3	2	1	1	1	1

Determining formal charge

Valences in the chart above represent common forms of these elements in organic compounds.

Many elements (e.g., chlorine, bromine, iodine) have several valences in different inorganic compounds.

If the number of covalent bonds to an atom is higher than its typical valence, it carries a positive formal charge.

If the number of covalent bonds to an atom is lower than its typical valence, it carries a negative formal charge.

Step-by-step approach to determine the formal charges of each atom in a molecule of nitric acid (HNO_3), using the structure displayed below.

$$HO-\overset{\overset{\displaystyle O}{\|}}{\underset{}{N^+}}-O^-$$

Formal charge for nitric acid (HNO₃) is shown on the atoms

Formal charge on **H**

- Hydrogen shares two electrons with oxygen.
- Assign one electron to H and one to O.
- Hydrogen has one valence electron, and $1 - 1 = 0$.
- Therefore, the formal charge of H in nitric acid is zero.

Formal charge on the **O** bonded to N and H

- Oxygen has four electrons in covalent bonds (2 with N, 2 with H).
- Assign two of these four electrons to O.
- Oxygen has two unshared pairs; assign four electrons to the oxygen atom.

- Total number of electrons assigned to O is $2 + 4 = 6$.

- Oxygen has six valence electrons, and $6 - 6 = 0$.

- Formal charge of O bonded to H and N in nitric acid is zero.

Formal charge on the **double-bonded O**

- Oxygen has four electrons in covalent bonds with N.

- Assign two of these four electrons to O.

- Oxygen has two unshared pairs; assign four electrons to O.

- Total number of electrons assigned to O is $2 + 4 = 6$.

- Oxygen has six valence electrons, and $6 - 6 = 0$.

- Formal charge of the double-bonded O in nitric acid is 0.

Formal charge on the **single-bonded O**

- O has two electrons in a covalent bond.

- Assign one of those electrons to O.

- O has three unshared pairs. Assign six electrons to O.

- The total number of electrons assigned to O is $1 + 6 = 7$.

- Oxygen has six valence electrons, $6 - 7 = -1$.

- Typical charge of single-bonded O in nitric acid is -1.

Formal charge of **N**

- N has eight electrons in covalent bonds.

- Assign four of those electrons to N.

- Nitrogen has five valence electrons, $5 - 4 = 1$.

- Therefore, the formal charge of the N in nitric acid is $+1$.

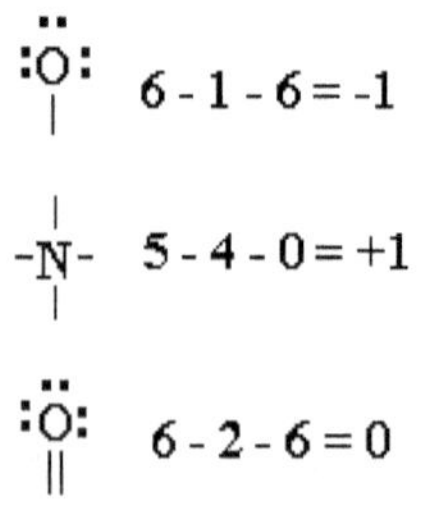

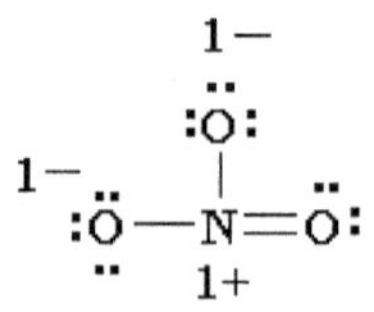

If atoms in a molecule have a formal charge, it does not necessarily mean that the entire molecule has a charge.

In the ozone structure (O_3), the central oxygen atom has three bonds and is positively charged.

The right-hand oxygen has a single bond and is negatively charged.

Overall charge of the ozone molecule is, therefore, zero.

ozone   *nitromethane*   *azide ion*

Nitromethane (CH_3NO_2) has a positively charged nitrogen and a negatively charged oxygen; the total molecular charge is again zero.

Azide (N_3^-) anion has two negatively charged nitrogens (2 bonds each) and one positively charged nitrogen (four bonds), and the total charge is minus one.

Notes for active learning

Notes for active learning

Resonance

Resonance structures

When more than one relatively stable structure exists, it forms *resonance structures*.

Resonance is the averaging of electron distribution over two or more hypothetical contributing structures to produce a hybrid electronic structure.

Resonance structures describe molecules that may contain fractional bonds and charges.

Resonance structures must have the same number of paired and unpaired electrons.

No atoms change position within the common structural framework, nor do they break existing chemical bonds.

Electrons (not atoms, as is the case for isomers) move. Visualize the molecule as quickly "shifting" between its resonance structures and intermittently existing in several configurations.

However, the molecule spends more time in the most stable resonance form.

More accurately, the molecular structure is a "combination" (or a weighted average) of its resonance structures, favoring the most stable resonance structures.

In a molecule with a single and double-bond resonance structure, the bond length is between a single-bond and a double-bond length.

$$O = S - O \qquad O - S - O \qquad O - S = O$$
$$\ \ \ \ \ | \qquad\qquad\quad || \qquad\qquad\quad\ \ |$$
$$\ \ \ \ O \qquad\qquad\quad O \qquad\qquad\quad\ \ O$$

Resonance structures for SO_3^{2-} (single-bonded O with negative charge). The images above do not include the formal charge or the double-headed arrows to show the individual contributing resonance structures.

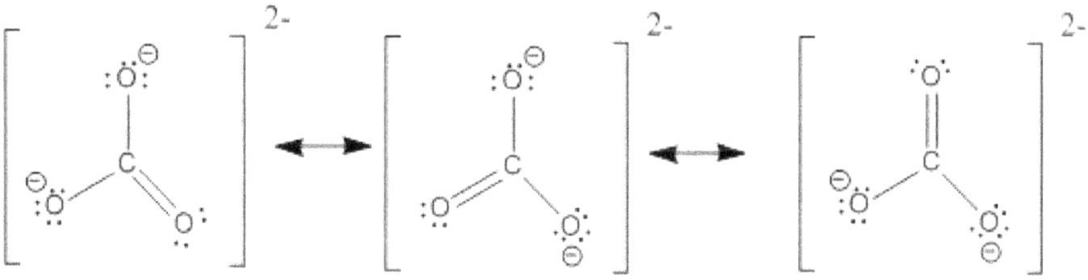

Three contributing resonance structures for CO_3^{2-}

Double-headed arrow seen in the image above for CO_3^{2-} acts as a notation symbol for resonance structures contributing to the hybrid structure. The principle of resonance is useful in rationalizing the chemical behavior of compounds.

Electron delocalization provided by resonance significantly enhances the molecule's stability, lowering its potential energy (PE).

Determine the most stable resonance structure of a molecule when a stable molecule's properties are known.

In stable molecules, the octet is satisfied in each atom (aside from hydrogen and exceptions, like boron).

Neutral molecules are more stable when the formal charges on atoms are distributed accordingly (i.e., a + charge on a positive atom and a – charge on electronegative atoms).

Resonance structure for formaldehyde (CH$_2$O)

In the example above, the neutral molecule is the most stable.

The preferred charge distribution has a positive charge on the less electronegative atom (carbon) and a negative charge on the more electronegative atom (oxygen).

The middle formula represents a more reasonable and stable structure than the one on the right.

Resonance hybrids

Stable structures indicate that the molecule spends most of its time in the resonance structure(s) over the others.

If the double bond is broken heterolytically (both electrons going to one of the bonding atoms), formal charge pairs result, as shown in the other two structures.

Since the middle, the charge-separated contributor has an electron-deficient carbon atom; this explains the tendency of electron donors (nucleophiles) to bond at this site.

The application of resonance to this example requires a weighted average of these structures.

Consequently, if one structure has greater stability than the others, the hybrid closely resembles it electronically and energetically.

Resonance hybrid is exceptionally stable, featuring two or more identical low-energy structures.

Examples of low-energy molecules stabilized by resonance hybrid structures include sulfur dioxide (SO_2) and nitric acid (HNO_3).

Three resonance forms of sulfur dioxide (SO_2); the central structure is neutral and therefore the greatest contributor (most molecules exist in this state)

Two resonance forms of nitric acid (HNO_3); structures are equal in energy with nitrogen positive, and the negative charge shared by two oxygens

Three considerations for drawing resonance structures

1. The number of covalent bonds in a structure - the greater the bonding, the more essential and stable the contributing structure.

2. Formal charge separation: *charge separation* decreases the stability and importance of the contributing structure.

3. Evaluate electronegativity of charge-bearing atoms and charge density.

    Positive charge is accommodated on atoms of low electronegativity.

    Negative charge is typically found on highly electronegative atoms.

Lewis acids and bases

The number of free electron pairs can categorize molecules:

   Lewis *acids accept* electron pairs.

   Lewis *bases donate* electron pairs.

Acids have vacant orbitals and do not have lone pairs on the central atom (e.g., BF_3), while Lewis bases have lone pair electrons (e.g., NH_3).

$$A^+ \ + \ {}^-B \ \longrightarrow \ A{-}B$$

Lewis acid (A): electron acceptor   Lewis base (B): electron donor

$$
\underset{\text{Acid}}{
\begin{array}{c}
\text{F} \\
| \\
\text{F}-\text{B} \\
| \\
\text{F}
\end{array}}
\; + \;
\underset{\text{Base}}{
\begin{array}{c}
\text{H} \\
| \\
{:}\text{N}-\text{H} \\
| \\
\text{H}
\end{array}}
\;\longrightarrow\;
\begin{array}{c}
\text{F} \quad \text{H} \\
| \qquad | \\
\text{F}-\text{B}-\text{N}-\text{H} \\
| \qquad | \\
\text{F} \quad \text{H}
\end{array}
$$

The product would have a formal negative charge on boron
and a positive charge on nitrogen

Acids and bases are discussed in the *Acids & Bases* title of this series.

Partial Ionic Character

Electron sharing

Covalent bonds can be categorized into nonpolar covalent bonds and polar covalent bonds.

Nonpolar covalent bonds occur between atoms of the same element or atoms with similar electronegativities, where electrons are shared equally to achieve a more stable electron configuration.

Polar covalent bonds form between atoms with differences in electronegativity. These bonds have *partial ionic character* due to the unequal distribution of the electrons. One atom has a stronger attraction for the shared electrons, resulting in one end of the bond being partially negatively charged ($\delta-$), while the other is partially positively charged (δ^+).

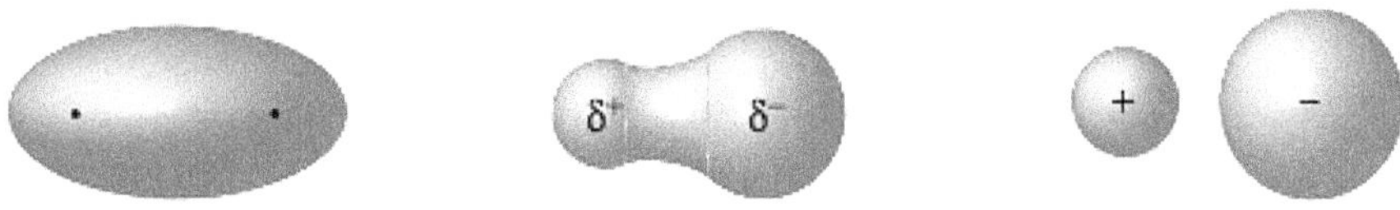

Three types of bonds: covalent (left), polar covalent (center), and ionic (right)

Left structure above illustrates a covalent bond between identical atoms (i.e., atoms with the same electronegativity).

Center structure above is a polar covalent bond between elements of moderately different electronegativities, with the notation for the molecule's resulting partial positive and partial negative ends.

Right figure above shows an ionic bond between elements with significant differences in electronegativities, typically a metal and a nonmetal. In ionic bonds, electrons from one atom are completely transferred to another, resulting in ions with opposite charges that are attracted to each other by electrostatic forces (e.g., in $NaCl$, sodium donates an electron to chlorine).

Nonpolar covalent bonds involve equal sharing of electrons, polar covalent bonds involve unequal sharing, and ionic bonds result from the complete transfer of electrons from one atom to the other. The difference in electronegativity between the atoms determines the type of bond.

Electronegativity for determining charge distribution

Electronegativity (EN) is an element's ability to attract electrons through bonding due to differing nuclear charges and shielding by inner electron shells; each element has different electronegativity.

The molecule's polarity is the difference in EN values of its atoms.

Linus Pauling, a Nobel laureate in chemistry (1954), elucidated the nature of the chemical bond and its application to the structure of complex molecules. Pauling established a quantitative scale of electronegativity values.

A larger number (e.g., F = 3.98) signifies a higher affinity for electrons.

H 2.20	Electronegativity values for some elements					
Li 0.98	**Be** 1.57	**B** 2.04	**C** 2.55	**N** 3.04	**O** 3.44	**F** 3.98
Na 0.90	**Mg** 1.31	**Al** 1.61	**Si** 1.90	**P** 2.19	**S** 2.58	**Cl** 3.16
K 0.82	**Ca** 1.00	**Ga** 1.81	**Ge** 2.01	**As** 2.18	**Se** 2.55	**Br** 2.96

For the periodic table, EN values increase from left to right, from bottom to top

Higher electronegativity values are towards the nonmetal side (right side) of the periodic table.

Fluorine has the highest electronegativity (3.98). The heavier alkali metals (e.g., potassium, rubidium, cesium) have lower electronegativity values.

Carbon has an electronegativity of 2.55, a value within the mid-range for electronegativity and is slightly more electronegative than hydrogen (2.20).

Of an atom's subatomic particles (protons, neutrons, electrons), only the electrons move within the atom.

When two atoms bond, most electrons congregate between the two nuclei.

The distribution depends on the electronegativity of the atoms.

Shared electrons in a bond are attracted to the more electronegative atom, shifting electron density toward the electronegative atom.

Nonpolar covalent bonds form between identical atoms as *homonuclear diatomic molecules*.

For example, in diatomic hydrogen (H_2), both atoms are identical; each has the same electron affinity (EA) value.

Thus, electrons are evenly distributed between the two hydrogens.

$$H : H \qquad H \; H$$

Polarity based on electronegativity

Hydrogen fluoride (HF) has atoms of different EN values – fluorine (3.98) is more electronegative than hydrogen (2.20). The electrons are more attracted to the fluorine nucleus and are closer to the F than H.

$$H \; : \; F \qquad H \quad F$$

Electron density for the H–F bond has the greatest density on the electronegative F

For example, suppose one atom in a molecule is more electronegative than another. This causes a higher concentration of electrons to locate on one side of the bond, and the molecule is polar (i.e., dipole present).

The molecule's *polarity* is proportional to the difference in the bonded atoms' electronegativity, shown below.

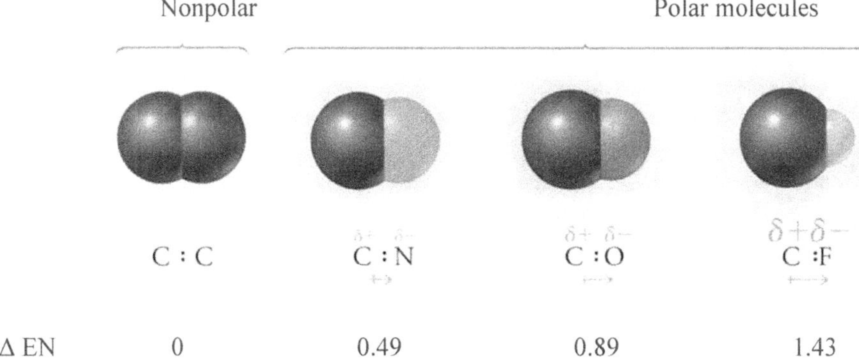

Lowercase Greek letter delta δ denotes partial charges on a polar molecule.

A plus or minus sign (i.e., (δ^+ or δ^-) indicates the partial positive or negative charges.

$$\delta^+ \; H\!-\!Cl \; \delta^-$$

Arrows show the direction of partial charges. The arrow points in the direction of electron movement, originating at the positively charged atom and pointing toward the negatively charged atom.

The partially positively charged $\delta+$ and negatively charged $\delta-$ ends are *dipoles*.

If the difference in electronegativity is high enough in a direction, the molecule acquires a net dipole moment (geometric polarity).

Dipole moment

Dipole moment measures polarity in a molecule, representing the difference in *electronegativity* values between the participating or bonded atoms.

Geometric shape of a molecule affects its *overall dipole moment.*

A molecule may have polar bonds, but if those bonds arrange to cancel the dipole moments, it has a zero net dipole and is nonpolar.

Molecule is nonpolar if:

1) It is symmetrical.

2) It has identical atoms surrounding the central atom.

3) Non-bonding electron pairs are evenly distributed and cancel (illustrated below).

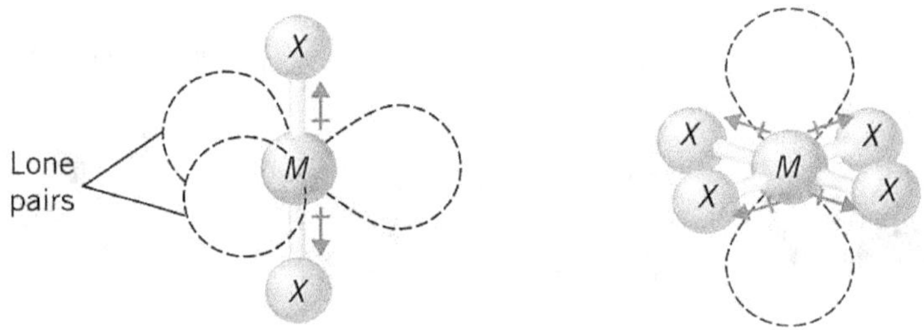

Symmetrical molecules have no net dipole, even with individual bond polarity

Dipole based on shape

In the right-hand linear configuration of the molecule below (bond angle 180°), the bond dipoles cancel, and the molecular dipole is zero.

Molecular dipole varies in size for bond angles ranging from 90° to 120°.

Molecular dipole is largest at the 90° configuration.

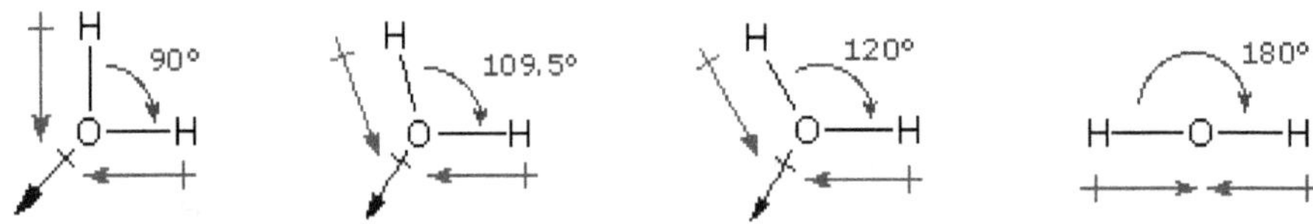

For example, boron trichloride (BCl_3) has three B–Cl bonds. B–Cl is a polar bond, but because the molecule is symmetrical and the charges are evenly distributed in a single plane, the charges cancel (shown below).

BCl₃ does not have non-bonding electron pairs, and the symmetrical BCl₃ molecule has a zero dipole moment.

$$Cl-B(Cl)(Cl)$$

Boron trichloride, BCl₃, is a symmetric molecule with a zero dipole moment

Configurations of methane (CH_4) and carbon dioxide (CO_2) may be deduced from their zero dipole moments.

Since the dipoles cancel, the configurations of these molecules must be tetrahedral and linear, respectively.

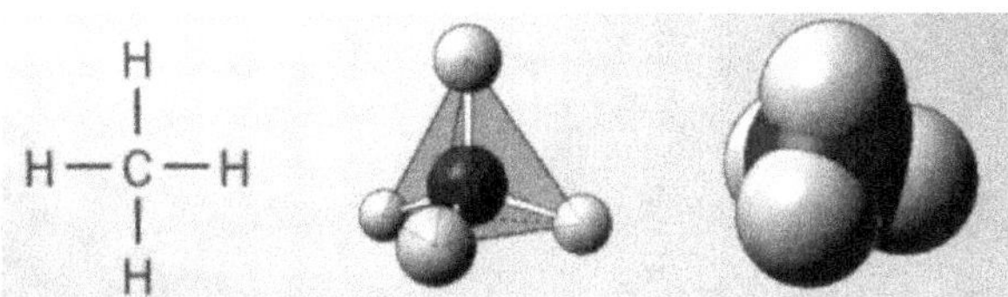

Methane (l to r) as simple drawing, ball-and-stick, and space-filling models

$$O=C=O$$

Carbon dioxide (CO_2) is a linear molecule and exhibits a net dipole of zero

Notes for active learning

Notes for active learning

Polarity

Molecular polarity

Methane CH_4 molecules (below) provide insight confirming its tetrahedral configuration.

Substituting one hydrogen atom with a chlorine atom yields the methyl chloride (CH_3Cl) molecule below.

Since the tetrahedral, square-planar, and square-pyramidal configurations have structurally equivalent hydrogen atoms, each gives a single substitution product.

One chlorine atom at the apex has a different electronegativity in the trigonal-pyramidal configuration than the three hydrogens at the pyramid base.

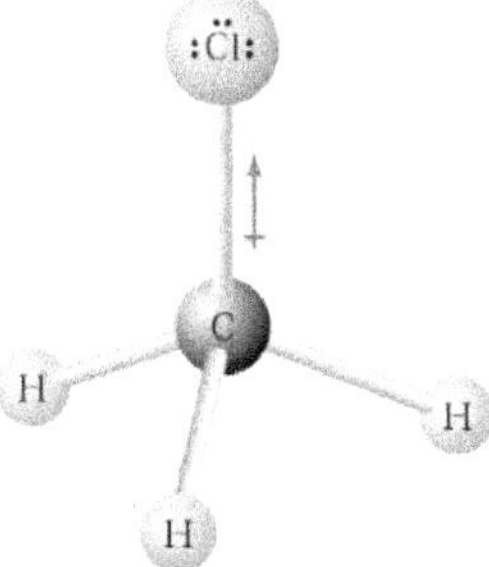

Methyl chloride (CH₃Cl; or chloromethane) with dipole pointing towards the single chlorine

Substitution should give two CH_3Cl compounds when hydrogen reacts.

Tetrahedral configuration of methane leads to a single dichloromethane CH_3Cl product below.

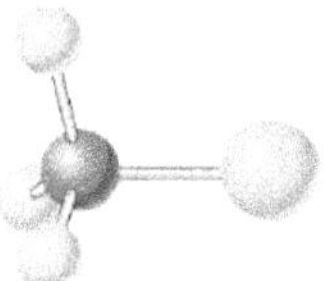

Methyl chloride, CH_3Cl, is a refrigerant used to manufacture synthetic rubber and silicones.

The *dipole moment* points toward the single electronegative chlorine atom.

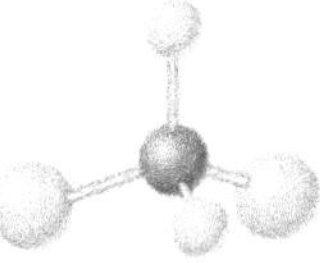

Dichloromethane (methyl chloride; above) CH_2Cl_2 is an industrial solvent used to remove paint and varnish. The *dipole moment* points toward the two electronegative chlorine atoms.

The vector is resolved between the two chlorines and away from the two hydrogens.

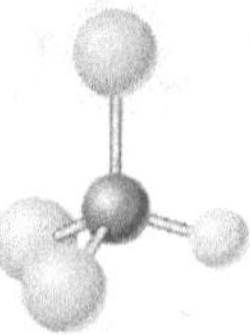

Trichloromethane (chloroform; above) $CHCl_3$ is a solvent to remove resins and adhesives.

Dipole moment points toward the three electronegative chlorine atoms.

The vector is resolved mathematically between the three Cl and away from H.

Most covalent compounds show some degree of local charge separation, resulting from differences in electronegativity between the atoms.

If a molecule has polar bonds arranged so that the bonds are in a specific direction, the molecule has a net dipole in that same direction. Some examples of polar molecules are PCl_3 and HCN.

Phosphorus trichloride (PCl_3) has a trigonal pyramidal molecular shape

The phosphorus trichloride (PCl_3) molecule has three chlorine atoms positioned on one side.

Since chlorine is highly electronegative, the electrons are attracted to the chlorine side, resulting in a dipole moment.

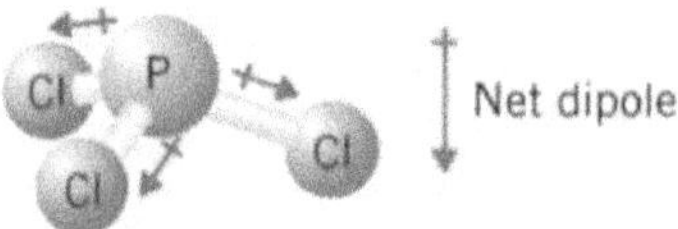

Phosphorus trichloride exhibits a net dipole because of the asymmetric pyramidal shape

Hydrogen cyanide (HCN) is a symmetrical molecule with one atom on each side of carbon.

Nitrogen is more electronegative than hydrogen; therefore, electrons tend to concentrate near the nitrogen.

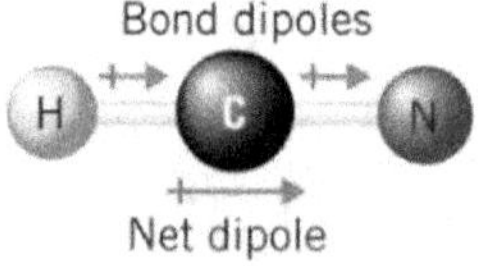

Polarity affects physical properties

Some physical properties (e.g., melting, boiling) are affected by molecular polarity.

Polar molecules have stronger intermolecular forces, resulting in higher boiling and melting points (as shown below).

Substance	Boiling Point (°C)
Polar	
Hydrogen fluoride, HF	20
Water, H_2O	100
Ammonia, NH_3	−33
Nonpolar	
Hydrogen, H_2	−253
Oxygen, O_2	−183
Nitrogen, N_2	−196
Boron trifluoride, BF_3	−100
Carbon dioxide, CO_2	−79

Boiling points of polar (high BP) and nonpolar (low BP) substances

A critical example of polarity is the water (H_2O) molecules (shown below).

The center oxygen of water has two pairs of non-bonding electrons, two hydrogens (one on each side), and is theoretically symmetrical if the two lone pairs are ignored.

VSEPR theory predicts that water has a bent shape, with the electron pairs arranged in a specific pattern (below).

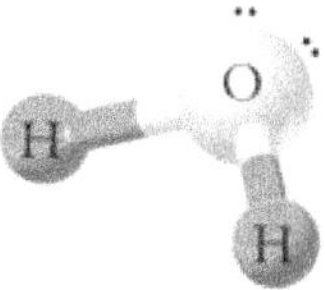

Water has two lone pairs. The dipole moment is indicated in the central structure.

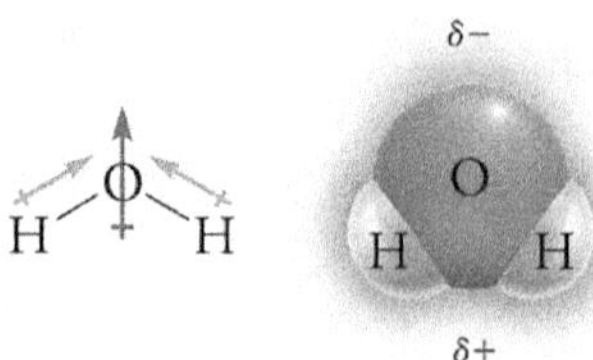

The space-filling water model with the molecule's partial negative and partial positive ends.

For water, the electron pairs are not evenly distributed; they are shifted downwards and concentrated on one side of the oxygen atom. That side has a partial negative charge, while the opposing side has a partial positive charge.

Therefore, the water molecule has a net dipole moment and is a polar molecule.

Polar water molecules exert strong intermolecular forces, known as *hydrogen bonds*.

Hydrogen bonds are a particular class of strong dipole-dipole interactions in which a hydrogen atom is bonded directly to one of the three most electronegative atoms (F, Cl, or N).

Notes for active learning

Notes for active learning

Notes for active learning

Practice Questions
&
Detailed Explanations

Page intentionally left blank

Practice Questions

Practice Set 1: Questions 1–20

---

1. What is the number of valence electrons in tin (Sn)?

   A. 14   **B.** 8   **C.** 2   **D.** 4   **E.** 5

2. Unhybridized p orbitals participate in π bonds as double and triple bonds. How many distinct and degenerate p orbitals exist in the second electron shell, where n = 2?

   A. 3   **B.** 2   **C.** 1   **D.** 0   **E.** 4

3. What is the number of valence electrons in a sulfite ion, SO_3^{2-}?

   A. 22   **C.** 26
   B. 24   **D.** 34
                                       E. none of the above

4. Which type of attractive forces occurs in molecules regardless of the atoms they possess?

   A. Dipole–ion interactions   **C.** Dipole–dipole attractions
   B. Ion ion interactions   **D.** Hydrogen bonding
                                         E. London dispersion forces

5. Given the structure of glucose, which statement explains the hydrogen bonding between glucose and water?

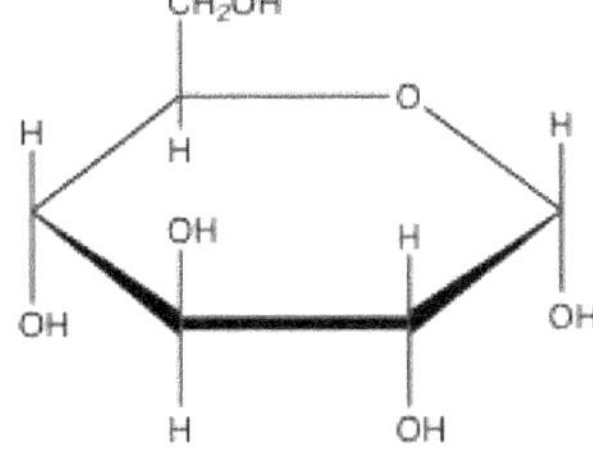

   A. Due to the cyclic structure of glucose, there is no H–bonding with water
   B. Each glucose molecule could H–bond with up to 17 water molecules
   C. H–bonds form, with water being the H–bond donor
   D. H–bonds form, with glucose being the H–bond donor
   E. Each glucose molecule could H–bond with up to 5 water molecules

6. How many valence electrons are in the Lewis dot structure of C_2H_6?

A. 2 **B.** 14 **C.** 6 **D.** 8 **E.** 12

7. The term for a bond where the electrons are shared unequally is:

A. ionic
B. nonpolar covalent

C. coordinate covalent
D. nonpolar ionic
E. polar covalent

8. What is the name for the weak forces of attraction between nonpolar molecules due to temporary dipoles between adjacent nonpolar molecules?

A. van der Waals forces
B. hydrophobic forces

C. hydrogen bonding forces
D. nonpolar covalent forces
E. hydrophilic forces

9. Nitrogen has five valence electrons; which of the following types of bonding is/are possible?

I. one single and one double bond
II. three single bonds
III. one triple bond

A. I only
B. II only

C. I and III only
D. I, II and III
E. none of the above

10. What is the number of valence electrons in antimony (Sb)?

A. 1 **B.** 2 **C.** 3 **D.** 4 **E.** 5

11. How many resonance structures, if any, can be drawn for a nitrite ion?

A. 1 **B.** 2 **C.** 3 **D.** 4 **E.** 5

12. What is the formula of the ammonium ion?

A. NH_4^-
B. N_4H^+

C. NH_4^{2-}
D. NH_4^{2+}
E. NH_4^+

13. What type of bond forms between oppositely charged ions?

A. dipole

B. covalent

C. London

D. induced dipole

E. ionic

14. Which of the following series of elements are arranged in the order of increasing electronegativity?

A. Fr, Mg, Si, O

B. Br, Cl, S, P

C. F, B, O, Li

D. Cl, S, Se, Te

E. Br, Mg, Si, N

15. The attraction due to London dispersion forces between molecules depends on which two factors?

A. Volatility and shape

B. Molar mass and volatility

C. Vapor pressure and size

D. Molar mass and shape

E. Molar mass and vapor pressure

16. Which of the substances below would have the largest dipole?

A. CO_2

B. SO_2

C. H_2O

D. CCl_4

E. CH_4

17. What is the name for the attraction between H_2O molecules?

A. adhesion

B. polarity

C. cohesion

D. van der Waals

E. hydrophilicity

18. Which compound contains only covalent bonds?

A. $HC_2H_3O_2$

B. NaCl

C. NH_4OH

D. $Ca_3(PO_4)_2$

E. LiF

19. The distance between two atomic nuclei in a chemical bond is determined by the:

A. size of the valence electrons

B. size of the nucleus

C. size of the protons

D. balance between the repulsion of the nuclei and the attraction of the nuclei for the bonding electrons

E. size of the neutrons

20. Carbonic acid has the chemical formula of H_2CO_3. The carbonate ion has the molecular formula of CO_3^{2-}. From the Lewis structure for the CO_3^{2-} ion, what is the number of reasonable resonance structures for the anion?

A. original structure only

B. 2

C. 3

D. 4

E. 5

Practice Set 2: Questions 21–40

---

21. Which of the following molecules would contain a dipole?

A. F–F

B. H–H

C. Cl–Cl

D. H–F

E. *trans*-dichloroethene

22. Which is the correct formula for the ionic compound formed between Ca and I?

A. Ca_3I_2 **B.** Ca_2I_3 **C.** CaI_2 **D.** Ca_2I **E.** Ca_3I_5

23. Which bonding is NOT possible for a carbon atom with four valence electrons?

A. 1 single and 1 triple bond

B. 1 double and 1 triple bond

C. 4 single bonds

D. 2 single and 1 double bond

E. 2 double bonds

24. During strenuous exercise, why does perspiration form droplets on a person's skin?

A. Ability of H_2O to dissipate heat

B. High specific heat of H_2O

C. Adhesive properties of H_2O

D. Cohesive properties of H_2O

E. High NaCl content of perspiration

25. If an ionic bond is stronger than a dipole–dipole interaction, why does water dissolve an ionic compound?

A. Ions do not overcome their interatomic attraction and therefore are not soluble

B. Ion–dipole interaction causes the ions to heat up and vibrate free of the crystal

C. Ionic bond is weakened by the ion–dipole interactions, and ionic repulsion ejects the ions from the crystal

D. Ion-dipole interactions of several water molecules aggregate with the ionic bond and dissociate it into the
solution

E. None of the above

26. Which one of these molecules can act as a hydrogen bond acceptor but not a donor?

A. CH_3NH_2

B. CH_3CO_2H

C. H_2O

D. C_2H_5OH

E. $CH_3–O–CH_3$

27. When NaCl dissolves in water, what is the force of attraction between Na^+ and H_2O?

A. ion–dipole

B. hydrogen bonding

C. ion–ion

D. dipole–dipole

E. van der Waals

28. Based on the Lewis structure, how many polar and nonpolar bonds are present in H_2CO?

A. 3 polar bonds and 0 nonpolar bonds

B. 2 polar bonds and 1 nonpolar bond

C. 1 polar bond and 2 nonpolar bonds

D. 0 polar bonds and 3 nonpolar bonds

E. 2 polar bonds and 2 nonpolar bonds

29. How many more electrons can fit within the valence shell of a hydrogen atom?

A. 1   **B.** 2   **C.** 7   **D.** 0   **E.** 3

30. Which element likely forms a cation with a +2 charge?

A. Na   **B.** S   **C.** Si   **D.** Mg   **E.** Br

31. Which of the statements accurately describes the structure of the ionic compound NaCl?

A. Alternating rows of Na^+ and Cl^- ions are present

B. Each ion present is surrounded by six ions of opposite charge

C. Alternating layers of Na and Cl atoms are present

D. Alternating layers of Na^+ and Cl^- ions are present

E. Repeating layers of Na^+ and Cl^- ions are present

32. What is the name for the force holding two atoms together in a chemical bond?

A. gravitational force

B. strong nuclear force

C. weak hydrophobic force

D. weak nuclear force

E. electrostatic force

33. Which of the following diatomic molecules contains the bond of greatest polarity?

A. CH_4 **B.** BrI   **C.** Cl–F   **D.** P_4 **E.** Te–F

34. Why does H_2O have an unusually high boiling point compared to H_2S?

A. Hydrogen bonding
B. Van der Waals forces
C. H_2O molecules pack more closely than H_2S
D. Covalent bonds are stronger in H_2O
E. This is a false statement because H_2O has a similar boiling point to H_2S

35. What is the shape of a molecule in which the central atom has 2 bonding electron pairs and 2 nonbonding electron pairs?

A. trigonal planar
B. trigonal pyramidal

C. linear
D. bent
E. tetrahedral

36. Which of the following represents the breaking of a noncovalent interaction?

A. Ionization of water
B. Decomposition of hydrogen peroxide

C. Hydrolysis of an ester
D. Dissolving of salt crystals
E. None of the above

37. Which pair of elements is most likely to form an ionic compound when reacted?

A. C and Cl
B. K and I

C. Ga and Si
D. Fe and Mg
E. H and O

38. Which of the following statements concerning coordinate covalent bonds is correct?

A. Once formed, they are indistinguishable from any other covalent bond
B. They are single bonds
C. One of the atoms involved must be a metal and the other a nonmetal
D. Both atoms involved in the bond contribute an equal number of electrons to the bond
E. The bond is formed between two Lewis bases

39. The greatest dipole moment within a bond is when:

A. both bonding elements have a low electronegativity

B. one bonding element has a high electronegativity, and the other has a low electronegativity

C. both bonding elements have a high electronegativity

D. one bonding element has a high electronegativity, and the other has a moderate electronegativity

E. both bonding elements have moderate electronegativity

40. Which of the following must occur for an atom to obtain the noble gas configuration?

A. lose, gain or share an electron

B. lose or gain an electron

C. lose an electron

D. share an electron

E. share or gain an electron

Practice Set 3: Questions 41–60

41. Based on the Lewis structure for hydrogen peroxide, H_2O_2, how many polar bonds and nonpolar bonds are present?

A. 3 polar and 0 nonpolar bonds

B. 2 polar and 2 nonpolar bonds

C. 1 polar and 2 nonpolar bonds

D. 0 polar and 3 nonpolar bonds

E. 2 polar and 1 nonpolar bond

42. To form an octet, an atom of selenium must:

A. gain 2 electrons

B. lose 2 electrons

C. gain 6 electrons

D. lose 6 electrons

E. gain 4 electrons

43. Which of the following is an example of a chemical reaction?

A. two solids mix to form a heterogeneous mixture

B. two liquids mix to form a homogeneous mixture

C. one or more new compounds are formed by rearranging atoms

D. a new element is formed by rearranging nucleons

E. a liquid undergoes a phase change and produces a solid

44. Which compound is NOT correctly matched with the predominant intermolecular force associated with that compound in the liquid state?

Compound	Intermolecular force
A. CH_3OH	hydrogen bonding
B. HF	hydrogen bonding
C. Cl_2O	dipole–dipole interactions
D. HBr	van der Waals interactions
E. CH_4	van der Waals interactions

45. Which element forms an ion with the greatest positive charge?

    A. Mg   **B.** Ca   **C.** Al   **D.** Na   **E.** Rb

46. What is the difference between a dipole–dipole and an ion–dipole interaction?

    A. One interaction involves dipole attraction between neutral molecules, while the other involves dipole interactions with ions

    B. One interaction involves ionic molecules interacting with other ionic molecules, while the other deals with polar molecules

    C. One interaction involves salts and water, while the other does not involve water

    D. One interaction involves hydrogen bonding, while the others do not

    E. None of the above

47. Which is a true statement about H_2O as it begins to freeze?

    A. Hydrogen bonds break

    B. Number of hydrogen bonds decreases

    C. Covalent bond strength increases

    D. Molecules move closer together

    E. Number of hydrogen bonds increases

48. In the process of forming sodium nitride (Na_3N) from its elements, what happens to the electrons of each sodium and the electrons of a nitrogen atom, respectively?

    A. one lost; three gained   **C.** three lost; one gained

    B. three lost; three gained   **D.** one lost; two gained

                                      E. two lost; three gained

49. Which species below has the least number of valence electrons in its Lewis symbol?

    A. S^{2-} **B.** Ga^+ **C.** Ar^+ **D.** Mg^{2+} **E.** F^-

50. Which of the following occur(s) naturally as nonpolar diatomic molecules?

    I. sulfur   II. chlorine   III. argon

    A. I only   **C.** I and III only

    B. II only   **D.** I and II only

                          E. I, II and III

51. In a chemical reaction, the bonds being formed are:

A. more energetic than the ones broken

B. less energetic than the ones broken

C. the same as the bonds broken

D. different from the ones broken

E. none of the above

52. The charge on a sulfide ion is:

A. +1
B. +2
C. 0
D. –2
E. –3

53. Which formula for an ionic compound is NOT correct?

A. $Al_2(CO_3)_3$

B. Li_2SO_4

C. Na_2S

D. $MgHCO_3$

E. K_2O

54. What term best describes the smallest whole number repeating ratio of ions in an ionic compound?

A. lattice

B. unit cell

C. formula unit

D. covalent unit

E. ionic unit

55. In the nitrogen monoxide molecule, the dipole moment is 0.16 D, and the bond length is 115 pm. What is the sign and magnitude of the charge on the oxygen atom? (Use the conversion factor of $1 \text{ D} = 3.34 \times 10^{-30}$ C·m and the charge of 1 electron $= 1.602 \times 10^{-19}$ C)

A. $-0.098 \ e$

B. $-0.71 \ e$

C. $-1.3 \ e$

D. $-0.029 \ e$

E. $+1.3 \ e$

56. From the electronegativity below, which single covalent bond is the most polar?

Element:	H	C	N	O
Electronegativity	2.1	2.5	3.0	3.5

A. O–C

B. O–N

C. N–C

D. C–H

E. C–C

57. Which of the following pairs is NOT correctly matched?

Formula	Molecular Geometry		Formula	Molecular Geometry
A. CH_4	tetrahedral		**C.** PCl_3	trigonal planar
B. OF_2	bent		**D.** Cl_2CO	trigonal planar
			E. $^+CH_3$	trigonal planar

58. Why are adjacent water molecules attracted to each other?

A. Ionic bonding between the hydrogens of H_2O

B. Covalent bonding between adjacent oxygens

C. Electrostatic attraction between the H of one H_2O and the O of another

D. Covalent bonding between the H of one H_2O and the O of another

E. Electrostatic attraction between the O of one H_2O and the O of another

59. What is the chemical formula for a compound that contains K^+ and CO_3^{2-} ions?

A. $K(CO_3)_3$

B. $K_3(CO_3)_2$

C. $K_3(CO_3)_3$

D. KCO_3

E. K_2CO_3

60. Under what conditions is graphite converted to diamond?

A. low temperature, high-pressure

B. high temperature, low-pressure

C. high temperature, high-pressure

D. low temperature, low-pressure

E. none of the above

Practice Set 4: Questions 61–80

61. Which of the following molecules is a Lewis acid?

A. NO_3^-

B. NH_3

C. NH_4^+

D. CH_3COOH

E. BH_3

62. Which molecule(s) is/are most likely to show a dipole-dipole interaction?

I. $H–C{\equiv}C–H$ II. CH_4 III. CH_3SH IV. CH_3CH_2OH

A. I only

B. II only

C. III only

D. III and IV only

E. I and IV only

63. C=C, C=O, C=N and N=N bonds are observed in many organic compounds. However, C=S, C=P, C=Si, and other similar bonds are not often found. What is the most probable explanation for this observation?

A. The comparative sizes of $3p$ atomic orbitals make effective overlap between them less likely than between two $2p$ orbitals

B. S, P and Si do not undergo hybridization of orbitals

C. S, P and Si do not form π bonds due to the lack of occupied p orbitals in their ground state electron configurations

D. Carbon does not combine with elements below the second row of the periodic table

E. None of the above

64. Write the formula for the ionic compound formed from magnesium and sulfur:

A. Mg_2S

B. Mg_3S_2

C. MgS_2

D. MgS_3

E. MgS

65. Explain why chlorine, Cl_2, is a gas at room temperature, while bromine, Br_2, is a liquid.

    A. Bromine ions are held by ionic bonds

    B. Chlorine molecules are smaller and, therefore, pack tighter in their physical orientation

    C. Bromine atoms are larger, which causes the formation of a stronger induced dipole induced dipole attraction

    D. Chlorine atoms are larger, which causes the formation of a stronger induced dipole induced dipole attraction

    E. Bromine molecules are smaller and therefore pack tighter in their physical orientation

66. What is the major intermolecular force in $(CH_3)_2NH$?

    A. hydrogen bonding

    B. dipole–dipole attractions

    C. London dispersion forces

    D. ion–dipole attractions

    E. van der Waals forces

67. Which of the following correctly describes potassium oxide and its bond?

    A. It is a weak electrolyte with an ionic bond

    B. It is a strong electrolyte with an ionic bond

    C. It is a non-electrolyte with a covalent bond

    D. It is a strong electrolyte with a covalent bond

    E. It is a non-electrolyte with a hydrogen bond

68. An ion with an atomic number of 34 and 36 electrons has what charge?

    A. +2

    B. –36

    C. +34

    D. –2

    E. neutral

69. Based on the Lewis structure for $H_3C–NH_2$, the formal charge on N is:

    A. –1

    B. 0

    C. +2

    D. +1

    E. –2

70. The name of S^{2-} is:

A. sulfite ion

B. sulfide ion

C. sulfur

D. sulfate ion

E. sulfurous acid

71. An ionic bond forms between two atoms when:

A. protons are transferred from the nucleus of the nonmetal to the nucleus of the metal

B. each atom acquires a negative charge

C. electron pairs are shared

D. four electrons are shared

E. electrons are transferred from metallic to nonmetallic atoms

72. Which of the following pairings of ions is NOT consistent with the formula?

A. Co_2S_3 (Co^{3+} and S^{2-})

B. K_2O (K^+ and O^-)

C. Na_3P (Na^+ and P^{3-})

D. BaF_2 (Ba^{2+} and F^-)

E. KCl (K^+ and Cl^-)

73. Which of the following solids likely has the smallest exothermic lattice energy?

A. Al_2O_3

B. KF

C. NaCl

D. LiF

E. NaOH

74. In chlorine monoxide, chlorine has a charge of $+0.167$ e. If the bond length is 154 pm, what is the dipole moment of the molecule? (Use the conversion of 1 meter $= 1 \times 10^{-12}$ m/pm and the value of 1 electron $= 1.602 \times 10^{-19}$ C)

A. 2.30 D

B. 0.167 D

C. 1.24 D

D. 3.11 D

E. 1.65 D

75. All of the following are examples of polar molecules, EXCEPT:

A. H_2O

B. CCl_4

C. CH_2Cl_2

D. HF

E. CO

76. Which is the most likely noncovalent interaction between an alcohol and a carboxylic acid at pH = 10?

    A. formation of an anhydride bond

    B. dipole–dipole interaction

    C. dipole–charge interaction

    D. charge–charge interaction

    E. induced dipole–dipole interaction

77. Which electron geometry is characteristic of an sp^2 hybridized atom?

    A. linear

    B. tetrahedral

    C. trigonal bipyramidal

    D. bent

    E. trigonal planar

78. Which of the following statements about noble gases is NOT correct?

    A. They have very stable electron arrangements

    B. They are the most reactive of gases

    C. They exist in nature as individual atoms rather than in the molecular form

    D. They have 8 valence electrons

    E. They have a complete octet

79. How are intermolecular forces and solubility related?

    A. Solubility is a measure of how weak the intermolecular forces in the solute are

    B. Solubility is a measure of how strong a solvent's intermolecular forces are

    C. Solubility depends on the solute's ability to overcome the intermolecular forces in the solvent

    D. Solubility depends on the solvent's ability to overcome the intermolecular forces in a solute

    E. None of the above

80. In a bond between any two of the following atoms, the bonding electrons would be most strongly attracted to:

    A. I   **B.** He   **C.** Cs   **D.** Cl   **E.** Fr

Answer Key & Detailed Explanations

Answer Key

1: D	11: B	21: D	31: B	41: E	51: D	61: E	71: E
2: A	12: E	22: C	32: E	42: A	52: D	62: D	72: B
3: C	13: E	23: B	33: E	43: C	53: D	63: A	73: C
4: E	14: A	24: D	34: A	44: D	54: C	64: E	74: C
5: B	15: D	25: D	35: D	45: C	55: D	65: C	75: B
6: B	16: C	26: E	36: D	46: A	56: A	66: A	76: C
7: E	17: C	27: A	37: B	47: E	57: C	67: B	77: E
8: A	18: A	28: C	38: A	48: A	58: C	68: D	78: B
9: D	19: D	29: A	39: B	49: D	59: E	69: B	79: D
10: E	20: C	30: D	40: A	50: B	60: C	70: B	80: D

Practice Set 1: Questions 1–20

1. D is correct.

The valence shell is an atom's outermost shell (i.e., highest principal quantum number n).

Valence electrons are electrons of the outermost electron shell that can participate in a chemical bond.

The number of valence electrons for an element can be determined by its group (i.e., vertical column) on the periodic table. Except for the transition metals (i.e., groups 3-12), the group number identifies how many valence electrons are associated with a particular element: elements of the same group have the same number of valence electrons.

2. A is correct.

Three degenerate p orbitals exist for an atom with an electron configuration in the second shell or higher. The first shell only has access to s orbitals.

The d orbitals become available from n = 3 (third shell).

3. C is correct.

The valence shell is an atom's outermost shell (i.e., highest principal quantum number n).

Valence electrons are electrons of the outermost electron shell that can participate in a chemical bond.

The number of valence electrons for an element can be determined by its group (i.e., vertical column) on the periodic table.

Except for the transition metals (i.e., groups 3-12), the group number identifies how many valence electrons are associated with a particular element: elements of the same group have the same number of valence electrons.

To find the number of valence electrons in a sulfite ion, SO_3^{2-}, add the valence electrons of each atom:

Sulfur = 6; Oxygen = $(6 \times 3) = 18$

Total = 24

This ion has a net charge of -2, which indicates that it has 2 extra electrons.

Therefore, the number of valence electrons would be $24 + 2 = 26$ electrons.

4. E is correct.

London dispersion forces result from the momentary flux of valence electrons and are present in compounds; they are the attractive forces that hold molecules together.

They are the weakest of the intermolecular forces, and their strength increases with increasing size (i.e., surface area contact) and polarity of the molecules involved.

5. B is correct.

Each hydroxyl group (alcohol or ~OH) has oxygen with 2 lone pairs and one attached hydrogen.

Therefore, each hydroxyl group can participate in 3 hydrogen bonds:

5 hydroxyl groups $\times$ 3 bonds = 15 hydrogen bonds.

The oxygen of the ether group (C–O–C) in the ring has 2 lone pairs for an additional 2 H–bonds.

6. B is correct.

The valence shell is an atom's outermost shell (i.e., highest principal quantum number n).

Valence electrons are electrons of the outermost electron shell that can participate in a chemical bond.

The number of valence electrons for an element can be determined by its group (i.e., vertical column) on the periodic table. Except for the transition metals (i.e., groups 3-12), the group number identifies how many valence electrons are associated with a particular element: elements of the same group have the same number of valence electrons.

Lewis dot structure for ethane

7. E is correct.

In covalent bonds, the electrons can be shared either equally or unequally.

Polar covalent bonded atoms are covalently bonded compounds that involve unequal sharing of electrons due to large electronegativity differences (Pauling units of 0.4 to 1.7) between the atoms.

An example of this is water, where there is a polar covalent bond between oxygen and hydrogen. Water is a polar molecule with the oxygen partial negative while the hydrogens are partial positive.

8. A is correct.

Van der Waals forces involve nonpolar (hydrophobic) molecules, such as hydrocarbons. The van der Waals force is the total of attractive or repulsive forces between molecules and therefore can be either attractive or repulsive. It can include the force between two permanent dipoles, the force between a permanent dipole and a temporary dipole, or the force between two temporary dipoles.

Hydrogens, bonded directly to F, O or N, participate in hydrogen bonds. The hydrogen is partial positive (i.e., delta plus or $\partial+$) due to the bond to these electronegative atoms. The lone pair of electrons on the F, O, or N interacts with the $\partial+$ hydrogen to form a hydrogen bond.

Hydrogen bonds are a type of dipole–dipole and are the strongest intermolecular forces (i.e., between molecules), followed by other types of dipole–dipole, dipole–induced dipole, and van der Waals forces (i.e., London dispersion).

9. D is correct.

Representative structures include:

$HNCH_2$: one single and one double bond

NH_3 : three single bonds

HCN : one triple bond and one single bond

10. E is correct.

The valence shell is an atom's outermost shell (i.e., highest principal quantum number n).

Valence electrons are electrons of the outermost electron shell that can participate in a chemical bond.

H ·							He:
Li ·	·Be·	· B ·	· C ·	· N ·	: O ·	: F ·	:Ne:
Na·	·Mg·	·Al·	·Si·	· P ·	: S ·	:Cl·	:Ar:
K ·	·Ca·	·Ga·	·Ge·	·As·	:Se·	:Br·	: Kr:
Rb·	· Sr ·	·In·	·Sn·	·Sb·	:Te·	: I ·	:Xe:

Sample Lewis dot structures for some elements

The number of valence electrons for an element can be determined by its group (i.e., vertical column) on the periodic table.

Except for the transition metals (i.e., groups 3-12), the group number identifies how many valence electrons are associated with an element: elements of the same group have the same number of valence electrons.

11. B is correct.

The nitrite ion has the chemical formula NO_2^- with the negative charge distributed between the two oxygen atoms.

Two resonance structures of the nitrite ion

12. E is correct.

Nitrogen has 5 valence electrons. In ammonia, nitrogen has 3 bonds to hydrogen, and a lone pair remains on the central nitrogen atom.

The ammonium ion (NH_4^+) has 4 hydrogens, and the lone pair of the nitrogen has been used to bond to the H^+ that has been added to ammonia.

Formal charge is shown on the ammonium ion

13. E is correct.

In atoms or molecules that are ions, the number of electrons is not equal to the number of protons, which gives the molecule a charge (either positive or negative).

Ionic bonds form when the difference in electronegativity of atoms in a compound is greater than 1.7 Pauling units.

Ionic bonds involve transferring an electron from the electropositive element (along the left-hand column/group) to the electronegative element (along the right-hand column/groups) on the periodic table.

Oppositely charged ions are attracted to each other, and this attraction results in ionic bonds.

In an ionic bond, electron(s) are transferred from a metal to a nonmetal, giving both molecules a full valence shell and causing the molecules to associate with each other closely.

14. A is correct.

Electronegativity is a chemical property that describes an atom's tendency to attract electrons to itself.

The most electronegative atom is F, while the least electronegative atom is Fr. The trend for increasing electronegativity within the periodic table is up and toward the right (i.e., fluorine).

15. D is correct.

Hydrogen bonds are the strongest intermolecular forces (i.e., between molecules), followed by dipole–dipole, dipole–induced dipole, and van der Waals forces (i.e., London dispersion).

London dispersion forces are present in compounds; they are the attractive forces that hold molecules together. They are the weakest of the intermolecular forces, and their strength increases with the increasing size and polarity of the molecules involved.

16. C is correct.

For asymmetrical molecules (e.g., water is bent), use the geometry and difference of electronegativity values between the atoms.

For water, H (2.1 Pauling units) is more electropositive than O (3.5 Pauling units).

For sulfur dioxide, S (2.5 Pauling units) is more electropositive than O (3.5 Pauling units).

The reported dipole moment for water is 1.8 D compared to 1.6 D for sulfur dioxide.

CO_2 does not have any nonbonding electrons on the central carbon atom, and it is symmetrical, so it is non-polar and has zero dipole moment. CCl_4 is symmetrical as a tetrahedron and does not exhibit a net dipole.

CH_4 is symmetrical as a tetrahedron and does not exhibit a net dipole.

17. C is correct.

Cohesion is the property of like molecules sticking together. Hydrogen bonds join water molecules.

Adhesion is the attraction between unlike molecules (e.g., the meniscus observed from water molecules adhering to the graduated cylinder).

Polarity is the difference in electronegativity between bonded molecules. Polarity gives rise to the delta plus (on H) and the delta minus (on O), permitting hydrogen bonds to form between water molecules.

18. A is correct.

With little or no difference in electronegativity between the elements (i.e., Pauling units < 0.4), it is a nonpolar covalent bond, whereby the electrons are shared between the two bonding atoms.

Among the answer choices, the atoms of H, C and O are closest in magnitude for Pauling units for electronegativity.

19. D is correct.

Positively charged nuclei repel each other while each attracts the bonding electrons. These opposing forces reach equilibrium at the bond length.

20. C is correct.

In a carbonate ion (CO_3^{2-}), the carbon atom is bonded to 3 oxygen atoms.

Two of those bonds are single covalent bonds, and the oxygen atoms each have an extra (third) lone pair of electrons, which imparts a negative formal charge.

The remaining oxygen has a double bond with carbon.

Three resonance structures for the carbonate ion CO_3^{2-}

Notes for active learning

Notes for active learning

Practice Set 2: Questions 21–40

21. D is correct.

A dipole is a separation of full (or partial) positive and negative charges due to differences in the electronegativity of atoms.

22. C is correct.

Ionic bonds involve transferring an electron from the electropositive element (along the left-hand column/group) to the electronegative atom (along the right-hand column/groups) on the periodic table.

Ca is a group II element with 2 electrons in its valence shell.

I is a group VII element with 7 electrons in its valence shell.

Ca becomes Ca^{2+}, and each of the two electrons is joined to I, which becomes I^-.

23. B is correct.

An atom with 4 valence electrons can make a maximum of 4 bonds. 1 double and 1 triple bond equals 5 bonds, which exceeds the maximum allowable bonds.

24. D is correct.

Water molecules stick to each other (i.e., cohesion) due to the collective action of hydrogen bonds between individual water molecules. These hydrogen bonds are constantly breaking and reforming; many molecules are held together by these bonds. Water sticks to surfaces (i.e., adhesion) because of water's polarity.

On a highly smooth surface (e.g., glass), the water may form a thin film because the molecular forces between glass and water molecules (adhesive forces) are stronger than the cohesive forces between the water molecules.

25. D is correct.

Hydrogen bonds are the strongest intermolecular forces (i.e., between molecules), followed by dipole–dipole, dipole–induced dipole, and van der Waals forces (i.e., London dispersion).

When ionic compounds are dissolved, each ion is surrounded by more than one water molecule. The combined force ion–dipole interactions of several water molecules are stronger than a single ionic bond.

26. E is correct.

Hydrogens, bonded directly to F, O or N, participate in hydrogen bonds. The hydrogen is partial positive (i.e., delta plus or $\partial+$) due to the bond to these electronegative atoms. The lone pair of electrons on the F, O or N interacts with the $\partial+$ hydrogen to form a hydrogen bond.

The two lone pairs of electrons on the oxygen atom can each participate as a hydrogen bond acceptor. The molecule does not have hydrogen bonded directly to an electronegative atom (F, O or N) and cannot be a hydrogen bond donor.

27. A is correct.

Water is a bent molecule with a partial negative charge on the oxygen and a partial positive charge on each hydrogen. Therefore, the water molecule exists as a dipole. The Na^+ is an ion attracted to the partial negative charge on the oxygen in the water molecule.

28. C is correct.

The molecule H_2CO (formaldehyde) is shown below and has one C=O bond and two C–H bonds. The electronegative oxygen pulls electron density away from the carbon atom and creates a net dipole towards the oxygen, resulting in a polar covalent bond.

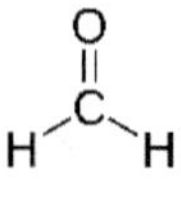

The electronegativity values of carbon and each hydrogen are similar and result in a covalent bond (i.e., about equal sharing of bonded electrons).

29. A is correct.

The valence shell is the outermost shell (i.e., highest principal quantum number n) of an atom.

Valence electrons are electrons of the outermost electron shell that can participate in a chemical bond.

The number of valence electrons for an element can be determined by its group (i.e., vertical column) on the periodic table.

Except for the transition metals (i.e., groups 3-12), the group number identifies how many valence electrons are associated with a particular element: elements of the same group have the same number of valence electrons.

30. D is correct.

Group IA elements (e.g., Li, Na, K) tend to lose 1 electron to achieve a complete octet to be cations with a +1 charge.

Group IIA elements (e.g., Mg, Ca) tend to lose 2 electrons to achieve a complete octet to be cations with a +2 charge.

Group VIIA elements (halogens such as F, Cl, Br and I) tend to gain 1 electron to achieve a complete octet to be anions with a −1 charge.

31. B is correct.

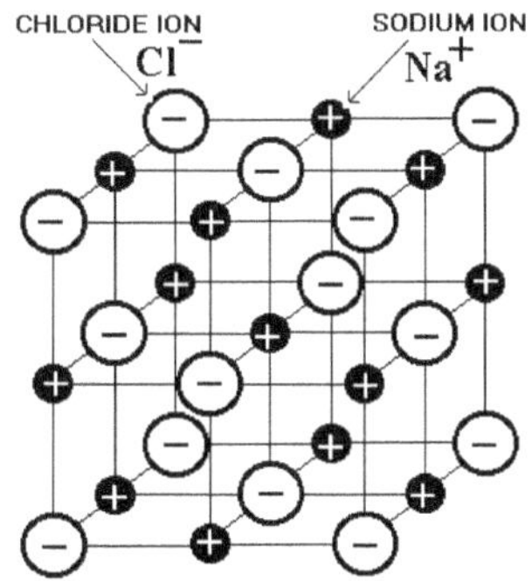

Lattice structure of sodium chloride

32. E is correct.

Each positively charged nuclei attracts the bonding electrons.

33. E is correct.

Electronegativity is a measure of how strongly an element attracts electrons within a bond.

Electronegativity is the relative attraction of the nucleus for bonding electrons. It increases from left to right (i.e., periods) and from bottom to top along a group (similar to the trend for ionization energy). The most electronegative atom is F, while the least electronegative atom is Fr.

The greater the difference in electronegativity between two atoms in a compound, the more polar of a bond these atoms form. The atom with the higher electronegativity is the partial (delta) negative end of the dipole.

34. A is correct.

Hydrogen bonds are the strongest intermolecular forces (i.e., between molecules), followed by dipole–dipole, dipole–induced dipole, and van der Waals forces (i.e., London dispersion).

Hydrogens, bonded directly to F, O or N, participate in hydrogen bonds. H–bonding is a polar interaction involving hydrogen forming bonds to the electronegative atoms F, O or N, which accounts for the high boiling points of water.

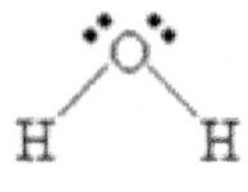

Molecular geometry of H₂S

The hydrogen is partial positive (i.e., delta plus or $\partial+$) due to the bond to these electronegative atoms. The lone pair of electrons on the F, O or N interacts with the $\partial+$ hydrogen to form a hydrogen bond.

Polar molecules have high boiling points because of polar interaction. H_2S is a polar molecule but does not form hydrogen bonds; it forms dipole–dipole interactions.

35. D is correct.

With 4 electron pairs, the starting shape is tetrahedral. After removing 2 of the 4 groups surrounding the central atom, the central atom has two groups bound (e.g., H_2O). The molecule has the hybridization (and original angles) from a tetrahedral shape, bent rather than linear.

36. D is correct.

Ionic bonds hold salt crystals. Salts are composed of cations and anions and are electrically neutral. When salts are dissolved in solution, they separate into their constituent ions by breaking noncovalent interactions.

Water: the hydrogen atoms in the molecule are covalently bonded to the oxygen atom.

Hydrogen peroxide: the molecule has one more oxygen atom than a water molecule and is held by covalent bonds.

Ester (RCOOR'): undergo hydrolysis and breaks covalent bonds to separate into its constituent carboxylic acid (RCOOH) and alcohol (ROH).

37. B is correct.

An ionic compound consists of a metal ion and a nonmetal ion.

Ionic bonds are formed between elements with an electronegativity difference greater than 1.7 Pauling units (e.g., a metal atom and a non-metal atom).

(K and I) form ionic bonds because K is metal and I is a nonmetal. (C and Cl) and (H and O) only have nonmetals, while (Fe and Mg) have only metals. (Ga and Si) has a metal (Ga) and a metalloid (Si); they *might* form weak ionic bonds.

38. A is correct.

A coordinate bond is a covalent bond (i.e., shared pair of electrons) in which both electrons come from the same atom.

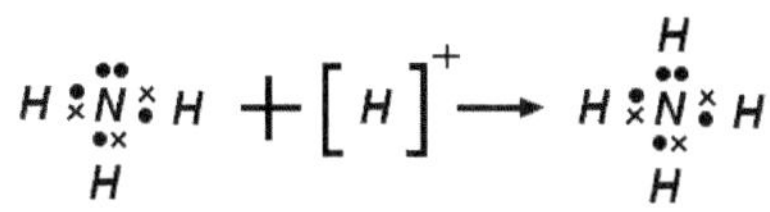

$$H \overset{\bullet\bullet}{\underset{\bullet\times}{\times N \times}} H \; + \; \left[H \right]^{+} \longrightarrow \; H \overset{H}{\underset{H}{\times \overset{\bullet\bullet}{N} \times}} H$$

The lone pair of the nitrogen is donated to form the fourth N–H bond

39. B is correct.

Dipole moment depends on the overall shape of the molecule, the length of the bond, and whether the electrons are pulled to one side of the bond (or the molecule overall).

For a large dipole moment, one element pulls electrons more strongly (i.e., differences in electronegativity).

Electronegativity measures how strongly an element attracts electrons within a bond.

Electronegativity is the relative attraction of the nucleus for bonding electrons. It increases from left to right (i.e., periods) and from bottom to top along a group (similar to the trend for ionization energy).

The most electronegative atom is F, while the least electronegative atom is Fr.

The greater the difference in electronegativity between two atoms in a compound, the more polar a bond these atoms form. The atom with the higher electronegativity is the partial (delta) negative end of the dipole.

40. A is correct.

The valence shell is the outermost shell (i.e., highest principal quantum number n) of an atom.

The noble gas configuration refers to eight electrons in the atom's outermost valence shell, referred to as a complete octet.

Depending on how many electrons it starts with, an atom may have to lose, gain or share an electron to obtain the noble gas configuration.

Notes for active learning

Practice Set 3: Questions 41–60

41. E is correct.

The bond between the oxygens is nonpolar, while the bonds between the oxygens and hydrogens are polar (due to the differences in electronegativity).

Line bond structure of H_2O_2 with lone pairs shown

42. A is correct.

The octet rule states that atoms of main-group elements tend to combine so that each atom has eight electrons in its valence shell. This occurs because electron arrangements involving eight valence electrons are extremely stable, as with noble gasses.

Selenium (Se) is in group VI and has 6 valence electrons. By gaining two electrons, it has a complete octet (i.e., stable).

43. C is correct.

When one or more new compounds are formed by rearranging atoms is an example of a chemical reaction.

44. D is correct.

HBr experiences dipole–dipole interactions due to the electronegativity difference resulting in a partial negative charge on bromine and a partial positive charge on hydrogen.

45. C is correct.

An ion is an atom (or a molecule) in which the number of electrons is not equal to the number of protons. Therefore, the atom (or molecule) has a net positive or negative electrical charge.

If a neutral atom loses one or more electrons, it has a net positive charge (i.e., cation).

If a neutral atom gains one or more electrons, it has a net negative charge (i.e., anion).

Aluminum is a group III atom with proportionally more protons per electron once the cation forms.

All other elements listed are from group I or II.

46. A is correct.

Dipole–dipole (e.g., $CH_3Cl...CH_3Cl$) attraction occurs between neutral molecules, while ion–dipole interaction involves dipole interactions with charged ions (e.g., $CH_3Cl..._{-}OOCCH_3$).

Hydrogen bonds are the strongest intermolecular forces (i.e., between molecules), followed by dipole–dipole, dipole–induced dipole, and van der Waals forces (i.e., London dispersion).

47. E is correct.

When more than two H_2O molecules are present (e.g., liquid water), more bonds (between 2 and 4) are possible because the oxygen of one water molecule has two lone pairs of electrons, each of which can form a hydrogen bond with hydrogen on another water molecule.

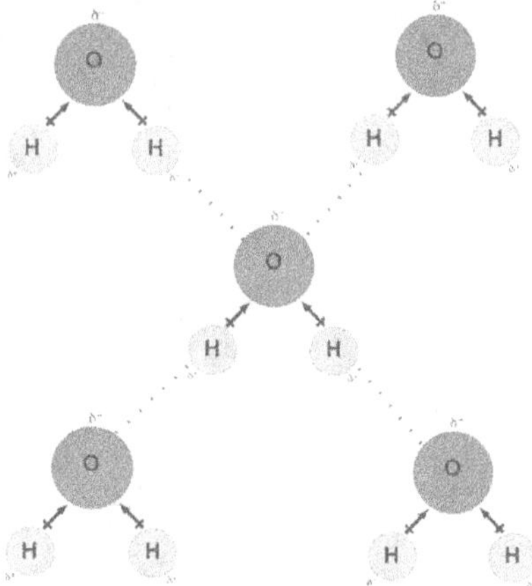

Ice with 4 hydrogen bonds between the water molecules

This bonding can repeat such that every water molecule is H–bonded with up to four other molecules (two through its two lone pairs of O, and two through its two hydrogen atoms).

Like other substances, it becomes increasingly dense when liquid water is cooled from room temperature. However, at approximately 4 °C (39 °F), water reaches its maximum density, and as it is cooled further, it expands and becomes less dense. This phenomenon is *negative thermal expansion* and is attributed to strong intermolecular interactions that are orientation-dependent.

The density of water is about 1 g/cm^3 and depends on the temperature. When frozen, the density of water is decreased by about 9%. This is due to the decrease in intermolecular vibrations, allowing water molecules to form stable hydrogen bonds with other water molecules around. As these hydrogen bonds form, molecules lock into positions similar to the hexagonal structure. Even though hydrogen bonds are shorter in the crystal than in the liquid, this position locking decreases the average coordination number of water molecules as the liquid reaches the solid phase.

48. A is correct.

The phrase "from its elements" in the question stem implies the need to create the formation reaction of the compound from its elements.

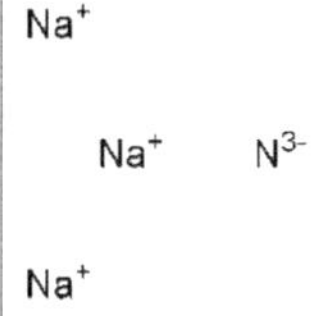

Chemical equation: $3\,Na^+ + N^{3-} \rightarrow Na_3N$

Each sodium loses one electron to form Na^+ ions.

Nitrogen gains 3 electrons to form the N^{3-} ion.

Schematic of an ionic compound

49. D is correct.

The valence shell is an atom's outermost shell (i.e., highest principal quantum number n).

Valence electrons are electrons of the outermost electron shell that can participate in a chemical bond.

The number of valence electrons for a neutral element can be determined by its group (i.e., vertical column) on the periodic table. Except for the transition metals (i.e., groups 3-12), the group number identifies how many valence electrons are associated with a particular element: elements of the same group have the same number of valence electrons.

H ·			· B ·	· C ·	· N̈ ·	: Ö ·	: F̈ ·	:N̈e:
Li ·	·Be·		· Al ·	· S̈i ·	· P̈ ·	: S̈ ·	:C̈l·	:Är:
Na·	·Mg·							
K ·	·Ca·		·G̈a·	·G̈e·	·Äs·	:S̈e·	:B̈r·	: K̈r:

Sample Lewis dot structures for some neutral elements

Lewis dot structures show the valence electrons of an individual atom as dots around the element's symbol. Non-valence electrons (i.e., inner shell electrons) are not represented in Lewis structures.

Mg^{2+} a total of ten electrons but zero valance (outer shell) electrons. The valence shell (i.e., 3s subshell) has zero electrons, and therefore there are no dots shown in the Lewis dot structure: Mg^{2+}

S^{2-} has eight valence electrons (same electronic configuration as Ar).

Ga^+ has two valence electrons (same electronic configuration as Ca).

Ar^+ has seven valence electrons (same electronic configuration as Cl).

F^- has eight valence electrons (same configuration as Ne). Lewis dot structure for F^-.

50. B is correct.

The octet rule states that atoms of main-group elements tend to combine so that each atom has eight electrons in its valence shell.

51. D is correct.

In a chemical reaction, the bonds being formed are different from the ones broken.

52. D is correct.

Hydrogen sulfide (i.e., sewer gas) is H_2S.

Hydrosulfide (bisulfide) ion is HS^-. Sulfide ion is S^{2-}

53. D is correct.

An ionic compound consists of a metal ion and a nonmetal ion.

Ionic bonds are formed between elements with an electronegativity difference greater than 1.7 Pauling units (e.g., a metal atom and a non-metal atom).

The charge on bicarbonate ion (HCO_3) is -1, while the charge on Mg is $+2$.

The proper formula is $Mg(HCO_3)_2$

54. C is correct.

Lattice: crystalline structure – consists of unit cells

Unit cell: smallest unit of solid crystalline pattern

Covalent unit and *ionic unit* are not valid terms.

55. D is correct.

Formula to calculate dipole moment:

$$\mu = qr$$

where μ is dipole moment (coulomb-meter or C·m), q is a charge (coulomb or C), and r is the radius (meter or m)

Convert the unit of dipole moment from Debye to C·m:

$$0.16 \times 3.34 \times 10^{-30} = 5.34 \times 10^{-31}\,C{\cdot}m$$

Convert the unit of radius to meter:

$$115\ pm \times 1 \times 10^{-12}\,m/pm = 115 \times 10^{-12}\,m$$

continued...

Rearrange the dipole moment equation to solve for q:

$$q = \mu / r$$

$$q = (5.34 \times 10^{-31}\,C{\cdot}m) / (115 \times 10^{-12}\,m)$$

$$q = 4.64 \times 10^{-21}\,C$$

Express the charge in terms of electron charge (e).

$$(4.64 \times 10^{-21}\,C) / (1.602 \times 10^{-19}\,C/e) = 0.029\ e$$

In this NO molecule, oxygen is the more electronegative atom.

Therefore, the charge experienced by the oxygen atom is negative: –0.029e.

56. A is correct.

The greater the difference in electronegativity between two atoms in a compound, the more polar a bond these atoms form. The atom with the higher electronegativity is the partial (delta) negative end of the dipole.

57. C is correct.

Hybridization of the central atom and the corresponding molecular geometry:

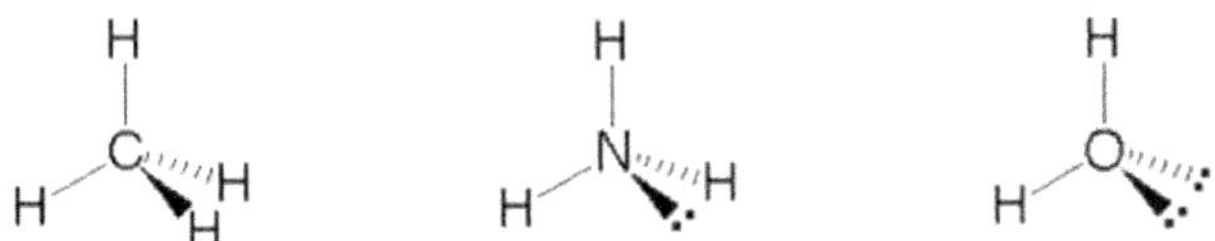

Examples of sp³ hybridized atoms with 4 substituents (CH4),
3 substituents (NH3) and two substituents (H2O)

<u>Hybridization - Shape</u>

sp – linear

sp^2 – trigonal planar

sp^3 – tetrahedral

sp^3d – trigonal bipyramid

58. C is correct.

Hydrogen bonds are the strongest intermolecular forces (i.e., between molecules), followed by dipole–dipole, dipole–induced dipole, and van der Waals forces (i.e., London dispersion).

Hydrogen is a very electropositive element, and O is an electronegative element. Therefore, these elements will be attracted to each other, both within the same water molecule and between water molecules.

Ionic or covalent bonding can only form within a molecule and not between adjacent molecules.

59. E is correct.

In a compound, the sum of ionic charges must equal zero.

For the charges of K^+ and CO_3^{2-} to even out, there have to be 2 K^+ ions for every CO_3^{2-} ion.

The formula of this balanced molecule is K_2CO_3.

60. C is correct.

The substantial difference between the two forms of carbon (i.e., graphite and diamonds) is mainly due to their crystal structure, hexagonal for graphite and cubic for diamond.

The conditions to convert graphite into diamond are high pressure and high temperature. That is why creating synthetic diamonds is time-consuming, energy-intensive, and expensive since carbon is forced to change its bonding structure

Practice Set 4: Questions 61–80

61. E is correct.

Lewis acids are *electron-pair acceptors*, whereas Lewis bases are electron-pair donors.

Boron has an atomic number of five and has a vacant $2p$ orbital to accept electrons.

62. D is correct.

CH_3SH experiences dipole–dipole interactions due to the electronegativity difference resulting in a partial negative charge on sulfur and a partial positive charge on hydrogen.

CH_3CH_2OH experiences dipole–dipole interactions (due to the electronegativity difference resulting in a partial negative charge on oxygen and a partial positive charge on hydrogen. This example is a dipole–dipole known as hydrogen bonding (when hydrogen is bonded directly to fluorine, oxygen, or nitrogen).

63. A is correct.

In carbon–carbon double bonds, there is an overlap of sp^2 orbitals and a p orbital on the adjacent carbon atoms.

The sp^2 orbitals overlap head-to-head as a *sigma* (σ) bond, whereas the p orbitals overlap sideways as a *pi* (π) bond.

Bond lengths and strengths (σ or π) depend on the size and shape of the atomic orbitals and the density to overlap effectively.

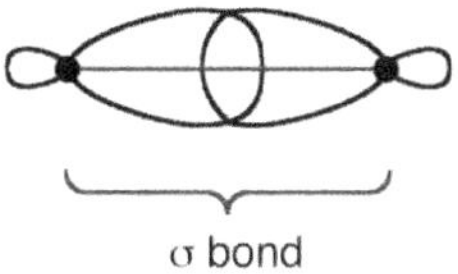

Sigma bond formation showing electron density along the internuclear axis

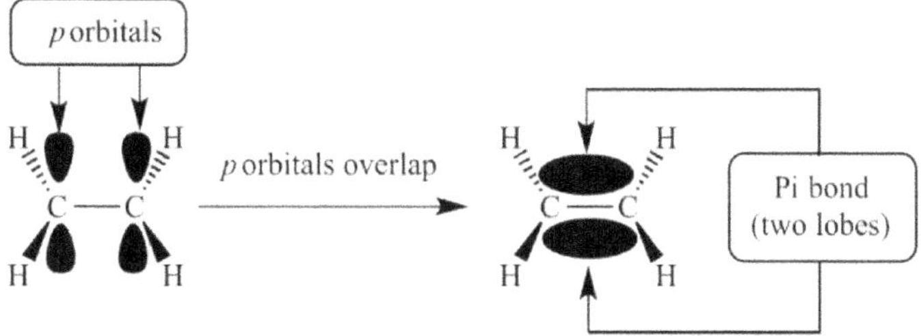

Two pi orbitals showing the pi bond formation during sideways overlap — note the absence of electron density (i.e., node) along the internuclear axis

The σ bonds are stronger than π bonds because head-to-head orbital overlap involves more shared electron density than sideways overlap. The σ bonds formed from two $2s$ orbitals are shorter than those formed from two $2p$ or two $3s$ orbitals.

Carbon, oxygen, and nitrogen are in the second period ($n = 2$), while sulfur (S), phosphorus (P), and silicon (Si) are in the third period. Therefore, S, P, and Si use $3p$ orbitals to form π bonds, while C, N, and O use $2p$ orbitals. The $3p$ orbitals are much larger than $2p$ orbitals, and therefore there is a reduced probability for an overlap of the $2p$ orbital of C and the $3p$ orbital of S, P, and Si.

B: S, P, and Si can hybridize, but these elements can combine s and p orbitals and (unlike C, O and N) have d orbitals.

C: S, P, and Si (ground-state electron configurations) have partially occupied p orbitals which form bonds.

D: carbon combines with elements below the second row of the periodic table. For example, carbon forms bonds with higher principal quantum number ($n > 2$) halogens (e.g., F, Cl, Br, and I).

64. E is correct.

The magnesium atom has a +2 charge, and the sulfur atom has a –2 charge.

65. C is correct.

Hydrogen bonds are the strongest intermolecular forces (i.e., between molecules), followed by dipole–dipole, dipole–induced dipole, and van der Waals forces (i.e., London dispersion).

Larger atoms have more electrons, which means they can exert stronger induced dipole forces than smaller ones.

Bromine is located under chlorine on the periodic table, which means it has more electrons and stronger induced dipole forces.

66. A is correct.

Hydrogen bonds are the strongest intermolecular forces (i.e., between molecules), followed by dipole–dipole, dipole–induced dipole, and van der Waals forces (i.e., London dispersion).

Hydrogens, bonded directly to F, O or N, participate in hydrogen bonds.

The hydrogen is partially positive (i.e., delta plus: $\partial+$) due to the bond to electronegative atoms.

The lone pair of electrons on the F, O or N interacts with the partial positive ($\partial+$) hydrogen to form a hydrogen bond.

When ~NH or ~OH groups are present in a molecule, they form hydrogen bonds with other molecules.

Hydrogen bonding is the strongest intermolecular force, and it is the major intermolecular force in this substance.

67. B is correct.

Potassium oxide (K_2O) contains metal and nonmetal, which form a salt.

Salts contain ionic bonds and dissociate in aqueous solutions and therefore are strong electrolytes.

A polyatomic (i.e., molecular) ion is a charged chemical species composed of two or more atoms covalently bonded or composed of a metal complex acting as a single unit.

An example is the hydroxide ion consisting of one oxygen atom and one hydrogen atom; hydroxide has a charge of -1.

A polyatomic ion does bond with other ions.

A polyatomic ion has various charges.

A polyatomic ion might contain only metals or nonmetals.

A polyatomic ion is not neutral.

oxidation state	-1	$+1$	$+3$	$+5$	$+7$
anion name	chloride	hypochlorite	chlorite	chlorate	perchlorate
formula	Cl^-	ClO^-	ClO_2^-	ClO_3^-	ClO_4^-

68. D is correct.

The valence shell is an atom's outermost shell (i.e., highest principal quantum number n).

Valence electrons are electrons of the outermost electron shell that can participate in a chemical bond.

69. B is correct.

Nitrogen (NH_3) is neutral with three bonds, negative with two bonds ($^-NH_2$), and positive with 4 bonds ($^+NH_4$)

Line bond structure of methylamine with the lone pair on nitrogen shown

70. B is correct.

Sulfide ion is S^{2-}.

Hydrogen sulfide (i.e., sewer gas) is H_2S.

Hydrosulfide (bisulfide) ion is HS^-.

Sulfite ion is the conjugate base of *bisulfite* and has the molecular formula of SO_3^{2-}

Sulfur (S) is a chemical element with an atomic number of 16.

The sulfate ion is a polyatomic anion (i.e., two or more atoms covalently bonded or metal complex) with the molecular formula of SO_4^{2-}

Sulfur acid (HSO_3^-) results from the combination of *sulfur dioxide* (SO_2) and H_2O.

71. E is correct.

Ionic bonds involve electrostatic interactions between oppositely charged atoms.

Electrons from the metallic atom are transferred to the nonmetallic atom, giving both atoms a full valence shell.

72. B is correct.

If the charge of O is -1, the proper formula of its compound with K ($+1$) should be KO.

73. C is correct.

The *lattice energy* of a crystalline solid is usually defined as the energy of formation of the crystal from infinitely-separated ions (i.e., an exothermic process and hence the negative sign).

Some older textbooks define lattice energy with the opposite sign. The older notation refers to the energy required to convert the crystal into infinitely separated gaseous ions in a vacuum (i.e., an endothermic process and the positive sign).

Lattice energy decreases downward for a group.

Lattice energy increases with charge.

Because NaCl and LiF contain charges of $+1$ and -1, they would have smaller lattice energy than the higher-charged ions on other options.

Na is lower than Li, and Cl is lower than F in their respective groups, so NaCl would have lower lattice energy than LiF.

The bond between ions of opposite charge is strongest when the ions are small.

The force of attraction between oppositely charged particles is directly proportional to the product of the charges on the two objects (q_1 and q_2) and inversely proportional to the square of the distance between the objects (r^2):

$$F = q_1 \times q_2 / r^2$$

Therefore, the strength of the bond between the ions of opposite charge in an ionic compound depends on the charges on the ions and the distance between the centers of the ions when they pack to form a crystal.

For NaCl, the lattice energy is the energy released by the reaction:

$$Na^+ (g) + Cl^- (g) \rightarrow NaCl (s), \text{ lattice energy} = -786 \text{ kJ/mol}$$

Other examples:

Al_2O_3, lattice energy $= -15,916$ kJ/mol

KF, lattice energy $= -821$ kJ/mol

LiF, lattice energy $= -1,036$ kJ/mol

NaOH, lattice energy $= -900$ kJ/mol

74. C is correct.

Formula to calculate *dipole moment*:

$$\mu = qr$$

where μ is dipole moment (coulomb per meter or C/m), q is a charge (coulomb or C), and r is the radius (meters or m)

Convert the unit of radius to meters:

$$154 \text{ pm} \times 1 \times 10^{-12} \text{ m/pm} = 154 \times 10^{-12} \text{ m}$$

Convert the charge to coulombs:

$$0.167 \, e \times 1.602 \times 10^{-19} \text{ C/}e^- = 2.68 \times 10^{-20} \text{ C}$$

Use these values to calculate the dipole moment:

$$\mu = qr$$

$$\mu = 2.68 \times 10^{-20} \text{ C} \times 154 \times 10^{-12} \text{ m}$$

$$\mu = 4.13 \times 10^{-30} \text{ C·m}$$

Finally, convert the dipole moment to Debye:

$$4.13 \times 10^{-30} \text{ C·m} / 3.34 \times 10^{-30} \text{ C·m·D}^{-1} = 1.24 \text{ D}$$

75. B is correct.

The greater the difference in electronegativity between two atoms in a compound, the more polar of a bond these atoms form. The atom with the higher electronegativity is the partial (delta) negative end of the dipole.

Although each C–Cl bond is very polar, the dipole moments of each of the four bonds in CCl_4 (carbon tetrachloride) cancel because the molecule is a symmetric tetrahedron.

76. C is correct.

The pK_a of the carboxylic acid is about 5 and is deprotonated (exists as an anion) at a pH of 10.

The pK_a of the alcohol is about 15 and is protonated (neutral) at a pH of 10.

77. E is correct.

Hybridization of the central atom and the corresponding electron geometry.

Hybridization	Electron groups	Bonding groups	Lone pairs	Approx. bond angles	Electron geometry	Molecular geometry
sp	2	2	0	180°	Linear	Linear
sp^2	3	3	0	120°	Trigonal planar	Trigonal planar
	3	2	1	<120°	Trigonal planar	Bent
sp^3	4	4	0	109.5°	Tetrahedral	Tetrahedral
	4	3	1	<109.5°	Tetrahedral	Trigonal pyramidal
	4	2	2	<<109.5°	Tetrahedral	Bent
sp^3d	5	5	0	120° (equatorial) 90° (axial)	Trigonal bipyramidal	Trigonal bipyramidal
	5	4	1	<120° (equatorial) <90° (axial)	Trigonal bipyramidal	Seesaw
	5	3	2	<90°	Trigonal bipyramidal	T-shaped
	5	2	3	180°	Trigonal bipyramidal	Linear
sp^3d^2	6	6	0	90°	Octahedral	Octahedral
	6	5	1	<90°	Octahedral	Square pyramidal
	6	4	2	90°	Octahedral	Square planar

78. B is correct.

The valence shell is an atom's outermost shell (i.e., highest principal quantum number n).

The octet rule states that atoms of main-group elements tend to combine so that each atom has eight electrons in its valence shell. This occurs because electron arrangements involving eight valence electrons are extremely stable, as with noble gasses.

79. D is correct.

Solubility depends on the solvent's ability to overcome the intermolecular forces in a solute.

80. D is correct.

Electronegativity is defined as the ability of an atom to attract electrons when it bonds with another atom.

The most electronegative atom is F, while the least electronegative atom is Fr. The trend for increasing electronegativity within the periodic table is up and toward the right (i.e., fluorine).

Chlorine has the highest electronegativity of the elements listed.

Electronegativity is a characteristic of non-metals.

Notes for active learning

Notes for active learning

APPENDIX

Periodic Table of the Elements

Key:
atomic number
Symbol
name
abridged standard atomic weight

1	2	3	4	5	6	7	8	9	10	11	12	13	14	15	16	17	18
1 **H** hydrogen 1.0080 ±0.0002																	2 **He** helium 4.0026 ±0.0001
3 **Li** lithium 6.94 ±0.06	4 **Be** beryllium 9.0122 ±0.0001											5 **B** boron 10.81 ±0.02	6 **C** carbon 12.011 ±0.002	7 **N** nitrogen 14.007 ±0.001	8 **O** oxygen 15.999 ±0.001	9 **F** fluorine 18.998 ±0.001	10 **Ne** neon 20.180 ±0.001
11 **Na** sodium 22.990 ±0.001	12 **Mg** magnesium 24.305 ±0.002											13 **Al** aluminium 26.982 ±0.001	14 **Si** silicon 28.085 ±0.001	15 **P** phosphorus 30.974 ±0.001	16 **S** sulfur 32.06 ±0.02	17 **Cl** chlorine 35.45 ±0.01	18 **Ar** argon 39.95 ±0.16
19 **K** potassium 39.098 ±0.001	20 **Ca** calcium 40.078 ±0.004	21 **Sc** scandium 44.956 ±0.001	22 **Ti** titanium 47.867 ±0.001	23 **V** vanadium 50.942 ±0.001	24 **Cr** chromium 51.996 ±0.001	25 **Mn** manganese 54.938 ±0.001	26 **Fe** iron 55.845 ±0.002	27 **Co** cobalt 58.933 ±0.001	28 **Ni** nickel 58.693 ±0.001	29 **Cu** copper 63.546 ±0.003	30 **Zn** zinc 65.38 ±0.02	31 **Ga** gallium 69.723 ±0.001	32 **Ge** germanium 72.630 ±0.008	33 **As** arsenic 74.922 ±0.001	34 **Se** selenium 78.971 ±0.008	35 **Br** bromine 79.904 ±0.003	36 **Kr** krypton 83.798 ±0.002
37 **Rb** rubidium 85.468 ±0.001	38 **Sr** strontium 87.62 ±0.01	39 **Y** yttrium 88.906 ±0.001	40 **Zr** zirconium 91.224 ±0.002	41 **Nb** niobium 92.906 ±0.001	42 **Mo** molybdenum 95.95 ±0.01	43 **Tc** technetium [97]	44 **Ru** ruthenium 101.07 ±0.02	45 **Rh** rhodium 102.91 ±0.01	46 **Pd** palladium 106.42 ±0.01	47 **Ag** silver 107.87 ±0.01	48 **Cd** cadmium 112.41 ±0.01	49 **In** indium 114.82 ±0.01	50 **Sn** tin 118.71 ±0.01	51 **Sb** antimony 121.76 ±0.01	52 **Te** tellurium 127.60 ±0.03	53 **I** iodine 126.90 ±0.01	54 **Xe** xenon 131.29 ±0.01
55 **Cs** caesium 132.91 ±0.01	56 **Ba** barium 137.33 ±0.01	57-71 lanthanoids	72 **Hf** hafnium 178.49 ±0.01	73 **Ta** tantalum 180.95 ±0.01	74 **W** tungsten 183.84 ±0.01	75 **Re** rhenium 186.21 ±0.01	76 **Os** osmium 190.23 ±0.03	77 **Ir** iridium 192.22 ±0.01	78 **Pt** platinum 195.08 ±0.02	79 **Au** gold 196.97 ±0.01	80 **Hg** mercury 200.59 ±0.01	81 **Tl** thallium 204.38 ±0.01	82 **Pb** lead 207.2 ±1.1	83 **Bi** bismuth 208.98 ±0.01	84 **Po** polonium [209]	85 **At** astatine [210]	86 **Rn** radon [222]
87 **Fr** francium [223]	88 **Ra** radium [226]	89-103 actinoids	104 **Rf** rutherfordium [267]	105 **Db** dubnium [268]	106 **Sg** seaborgium [269]	107 **Bh** bohrium [270]	108 **Hs** hassium [269]	109 **Mt** meitnerium [277]	110 **Ds** darmstadtium [281]	111 **Rg** roentgenium [282]	112 **Cn** copernicium [285]	113 **Nh** nihonium [286]	114 **Fl** flerovium [290]	115 **Mc** moscovium [290]	116 **Lv** livermorium [293]	117 **Ts** tennessine [294]	118 **Og** oganesson [294]

57 **La** lanthanum 138.91 ±0.01	58 **Ce** cerium 140.12 ±0.01	59 **Pr** praseodymium 140.91 ±0.01	60 **Nd** neodymium 144.24 ±0.01	61 **Pm** promethium [145]	62 **Sm** samarium 150.36 ±0.02	63 **Eu** europium 151.96 ±0.01	64 **Gd** gadolinium 157.25 ±0.03	65 **Tb** terbium 158.93 ±0.01	66 **Dy** dysprosium 162.50 ±0.01	67 **Ho** holmium 164.93 ±0.01	68 **Er** erbium 167.26 ±0.01	69 **Tm** thulium 168.93 ±0.01	70 **Yb** ytterbium 173.05 ±0.02	71 **Lu** lutetium 174.97 ±0.01
89 **Ac** actinium [227]	90 **Th** thorium 232.04 ±0.01	91 **Pa** protactinium 231.04 ±0.01	92 **U** uranium 238.03 ±0.01	93 **Np** neptunium [237]	94 **Pu** plutonium [244]	95 **Am** americium [243]	96 **Cm** curium [247]	97 **Bk** berkelium [247]	98 **Cf** californium [251]	99 **Es** einsteinium [252]	100 **Fm** fermium [257]	101 **Md** mendelevium [258]	102 **No** nobelium [259]	103 **Lr** lawrencium [262]

International Union of Pure and Applied Chemistry (IUPAC), 4 May 2022

Notes for active learning

Notes for active learning

Common Chemistry Equations

Throughout the test the following symbols have the definitions specified unless otherwise noted.

L, mL	=	liter(s), milliliter(s)		mm Hg	=	millimeters of mercury
g	=	gram(s)		J, kJ	=	joule(s), kilojoule(s)
nm	=	nanometer(s)		V	=	volt(s)
atm	=	atmosphere(s)		mol	=	mole(s)

ATOMIC STRUCTURE

$$E = h\nu$$
$$c = \lambda\nu$$

E = energy
ν = frequency
λ = wavelength

Planck's constant, $h = 6.626 \times 10^{-34}$ J s

Speed of light, $c = 2.998 \times 10^{8}$ m s^{-1}

Avogadro's number $= 6.022 \times 10^{23}$ mol^{-1}

Electron charge, $e = -1.602 \times 10^{-19}$ coulomb

EQUILIBRIUM

$$K_c = \frac{[\text{C}]^c[\text{D}]^d}{[\text{A}]^a[\text{B}]^b}, \text{ where } a\,\text{A} + b\,\text{B} \rightleftarrows c\,\text{C} + d\,\text{D}$$

$$K_p = \frac{(P_\text{C})^c(P_\text{D})^d}{(P_\text{A})^a(P_\text{B})^b}$$

$$K_a = \frac{[\text{H}^+][\text{A}^-]}{[\text{HA}]}$$

$$K_b = \frac{[\text{OH}^-][\text{HB}^+]}{[\text{B}]}$$

$$K_w = [\text{H}^+][\text{OH}^-] = 1.0 \times 10^{-14} \text{ at } 25°\text{C}$$
$$= K_a \times K_b$$

$$\text{pH} = -\log[\text{H}^+], \ \text{pOH} = -\log[\text{OH}^-]$$

$$14 = \text{pH} + \text{pOH}$$

$$\text{pH} = \text{p}K_a + \log\frac{[\text{A}^-]}{[\text{HA}]}$$

$$\text{p}K_a = -\log K_a, \ \text{p}K_b = -\log K_b$$

Equilibrium Constants

K_c (molar concentrations)
K_p (gas pressures)
K_a (weak acid)
K_b (weak base)
K_w (water)

KINETICS

$$\ln[\text{A}]_t - \ln[\text{A}]_0 = -kt$$

$$\frac{1}{[\text{A}]_t} - \frac{1}{[\text{A}]_0} = kt$$

$$t_{1/2} = \frac{0.693}{k}$$

k = rate constant
t = time
$t_{1/2}$ = half-life

GASES, LIQUIDS, AND SOLUTIONS

$$PV = nRT$$

$$P_A = P_{total} \times X_A, \text{ where } X_A = \frac{\text{moles A}}{\text{total moles}}$$

$$P_{total} = P_A + P_B + P_C + \ldots$$

$$n = \frac{m}{M}$$

$$K = {}^\circ C + 273$$

$$D = \frac{m}{V}$$

$$KE \text{ per molecule} = \frac{1}{2}mv^2$$

$$\text{Molarity}, M = \text{moles of solute per liter of solution}$$

$$A = abc$$

P = pressure
V = volume
T = temperature
n = number of moles
m = mass
M = molar mass
D = density
KE = kinetic energy
v = velocity
A = absorbance
a = molar absorptivity
b = path length
c = concentration

Gas constant, R = 8.314 J mol^{-1} K^{-1}
$\quad$ = 0.08206 L atm mol^{-1} K^{-1}
$\quad$ = 62.36 L torr mol^{-1} K^{-1}
1 atm = 760 mm Hg
$\quad$ = 760 torr
STP = $0.00\,^\circ$C and 10^5 Pa

THERMOCHEMISTRY/ ELECTROCHEMISTRY

$$q = mc\Delta T$$

$$\Delta S^\circ = \sum S^\circ \text{ products} - \sum S^\circ \text{ reactants}$$

$$\Delta H^\circ = \sum \Delta H_f^\circ \text{ products} - \sum \Delta H_f^\circ \text{ reactants}$$

$$\Delta G^\circ = \sum \Delta G_f^\circ \text{ products} - \sum \Delta G_f^\circ \text{ reactants}$$

$$\Delta G^\circ = \Delta H^\circ - T\Delta S^\circ$$

$$= -RT \ln K$$

$$= -nFE^\circ$$

$$I = \frac{q}{t}$$

q = heat
m = mass
c = specific heat capacity
T = temperature
S° = standard entropy
H° = standard enthalpy
G° = standard free energy
n = number of moles
E° = standard reduction potential
I = current (amperes)
q = charge (coulombs)
t = time (seconds)

Faraday's constant, F = 96,485 coulombs per mole
$\quad$ of electrons
1 volt = $\dfrac{1 \text{ joule}}{1 \text{ coulomb}}$

Annotated Glossary of Chemistry Terms

A

Absolute entropy (of a substance) – the increase in the entropy of a substance as it goes from a perfectly ordered crystalline form at 0 K (where its entropy is zero) to the temperature in question.

Absolute zero – the zero point on the absolute temperature scale; –273.15 °C or 0 K; theoretically, the temperature at which molecular motion ceases (i.e., the system does not emit or absorb energy, atoms at rest).

Absorption spectrum – spectrum associated with the absorption of electromagnetic radiation by atoms (or other species), resulting from transitions from lower to higher energy states.

Accuracy – the degree to which a value is close to the actual value; see *precision*.

Acid – a substance that produces H^+ (aq) ions in an aqueous solution and gives a pH of less than 7.0; strong acids ionize entirely or almost entirely in dilute aqueous solution; weak acids ionize only slightly. It turns litmus red.

Acid dissociation constant – an equilibrium constant for dissociating a weak acid.

Acid rain – rainwater with a pH of less than 5.7; caused by the gases NO_2 (vehicle exhaust fumes) and SO_2 (from burning fossil fuels) dissolving in the rain. It kills fish, wildlife, and trees and destroys buildings and lakes.

Acidic salt – contains an ionizable hydrogen atom; does not necessarily produce acidic solutions.

Actinides – the fifteen chemical elements that are between actinium (89) and lawrencium (103).

Activated complex – a structure formed by a collision between molecules in which new bonds form.

Activation energy – the amount of energy that reactants must absorb in their ground states to reach the transition state needed for a reaction to occur.

Active metal – a metal with low ionization energy that loses electrons readily to form cations.

Activity (of a component of ideal mixture) – a dimensionless quantity whose magnitude is equal to the molar concentration in an ideal solution; equal to partial pressure in an ideal gas mixture; 1 for pure solids or liquids.

Activity series – a listing of metals (and hydrogen) in order of decreasing activity.

Actual yield – the amount of a specified pure product obtained from a given reaction; see *theoretical yield*.

Addition reaction – a reaction in which two atoms or groups of atoms are added to a molecule, one on each side of a double or triple bond.

Adhesive forces – forces of attraction between a liquid and another surface.

Adsorption – the adhesion of a species onto the surfaces of particles.

Aeration – the mixing of air into a liquid or a solid.

Alcohol – a hydrocarbon derivative containing a ~OH group attached to a carbon atom, not in an aromatic ring.

Alkali metals – the elements of Group IA on the periodic table (e.g., Na, K, Rb).

Alkaline battery – a dry cell in which the electrolyte contains KOH.

Alkaline earth metals – group IIA metals on the periodic table; see *earth metals*.

Allomer – a substance that has a different composition from another but the same crystalline structure.

Allotropes – elements with different structures (therefore different forms), such as carbon (e.g., diamond, graphite, and fullerene).

Allotropic modifications (allotropes) – different forms of the same element in the same physical state.

Alloy – a mixture of metals. For example, bronze is an alloy formed from copper and tin.

Alloying – mixing metals with other substances (usually other metals) to modify their properties.

Alpha (α) particle – a helium nucleus; helium ion with 2+ charge; an assembly of two protons and two neutrons.

Amorphous solid – a non-crystalline solid with no well-defined ordered structure.

Ampere – unit of electrical current; one ampere equals one coulomb per second.

Amphiprotism – the ability of a substance to exhibit amphiprotic by accepting donated protons.

Amphoterism – the ability to react with both acids and bases; to act as either an acid or a base.

Amplitude – the maximum distance that medium particles carrying the wave move from their rest position.

Anion – a negative ion; an atom or group of atoms that has gained one or more electrons.

Anode – in a cathode ray tube, the positive electrode (electrode at which oxidation occurs); the positive side of a dry cell battery or a cell.

Antibonding orbital – a molecular orbital higher in energy than any of the atomic orbitals from which it is derived; lends instability to a molecule or ion when populated with electrons; denoted with star (*) superscript.

Artificial transmutation – an artificially induced nuclear reaction caused by the bombardment of a nucleus with subatomic particles or small nuclei.

Associated ions – short-lived species formed by the collision of dissolved ions of opposite charges.

Atmosphere – a unit of pressure; the pressure supports a column of mercury 760 mm high at 0 °C.

Atom – a chemical element in its smallest form; it comprises neutrons and protons within the nucleus and electrons circling the nucleus; it is the smallest part of an element that can exist.

Atomic mass unit (amu) – one-twelfth of the mass of an atom of the carbon-12 isotope; used for stating atomic and formula weights; known as a dalton.

Atomic number – represents an element corresponding with the number of protons within the nucleus; the number of protons in the nucleus of the atom.

Atomic orbital (*AO*) – a region or volume in space where the probability of finding electrons is highest.

Atomic radius – radius of an atom.

Atomic weight – weighted average of the masses of the constituent isotopes of an element; the relative masses of atoms of different elements.

Aufbau (or *building up*) principle – describes the order in which electrons fill orbitals in atoms.

Autoionization – an ionization reaction between identical molecules.

Avogadro's Law – equal volumes of gases contain the same number of molecules at the same temperature and pressure.

Avogadro's number (N_A) – the number (6.022×10^{23}) of atoms, molecules, or particles found in precisely 1 mole of a substance.

B

Background radiation – extraneous to an experiment; usually the low-level natural radiation from cosmic rays and trace radioactive substances present in the environment.

Band – a series of very closely spaced nearly continuous molecular orbitals that belong to the crystal as a whole.

Band of stability – band containing nonradioactive nuclides in a plot of neutrons *vs.* their atomic number.

Band theory of metals – the theory that accounts for the bonding and properties of metallic solids.

Barometer – a device used to measure the pressure in the atmosphere.

Base – a substance that produces $^-$OH (*aq*) ions in an aqueous solution; accepts a proton and has a high pH; strongly soluble bases are soluble in water and are entirely dissociated; weak bases ionize only slightly; a typical example of a base is sodium hydroxide ($NaOH$). It turns litmus blue.

Basic anhydride – the oxide of a metal that reacts with water to form a base.

Basic salt – a salt containing an ionizable OH group.

Beta (β) particle – an electron emitted from the nucleus when a neutron decays to a proton and an electron.

Binary acid – a binary compound in which H is bonded to one or more electronegative nonmetals.

Binary compound – consists of two elements; it may be ionic or covalent.

Binding energy (nuclear binding energy) – the energy equivalent ($E = mc^2$) of the mass deficiency of an atom (where E is the energy in joules, m is the mass in kilograms, and c is the speed of light in m/s^2).

Boiling – the phase transition of a liquid vaporizing.

Boiling point – the temperature at which the vapor pressure of a liquid is equal to the applied pressure; the *condensation point*.

Boiling point elevation – the increase in the boiling point of a solvent caused by the dissolution of a nonvolatile solute.

Bomb calorimeter – a device used to measure the heat transfer between a system and its surroundings at constant volume.

Bond – the attraction and repulsion between atoms and molecules is a cornerstone of chemistry.

Bond energy – the amount of energy necessary to break one mole of bonds in a substance, dissociating the substance in its gaseous state into atoms of its elements in the gaseous state.

Bond order – half the number of electrons in bonding orbitals minus half the electrons in antibonding orbitals.

Bonding orbital – a molecular orbit lower in energy than any of the atomic orbitals from which it is derived; lends stability to a molecule or ion when populated with electrons.

Bonding pair – pair of electrons involved in a covalent bond.

Boron hydrides – binary compounds of boron and hydrogen.

Born-Haber cycle – a series of reactions (and the accompanying enthalpy changes) which, when summed, represent the hypothetical one-step reaction by which elements in their standard states are converted into crystals of ionic compounds (and the accompanying enthalpy changes).

Boyle's Law – at a constant temperature, the volume occupied by a definite mass of a gas is inversely proportional to the applied pressure.

Breeder reactor – a nuclear reactor that produces more fissionable nuclear fuel than it consumes.

Brønsted-Lowrey acid – a chemical species that donates a proton.

Brønsted-Lowrey base – a chemical species that accepts a proton.

Buffer solution – resists change in pH; contains either a weak acid and a soluble ionic salt of the acid or a weak base and a soluble ionic salt of the base.

Buret – a piece of volumetric glassware, usually graduated in 0.1 mL intervals, used to deliver solutions for titrations in a quantitative (drop-like) manner; also spelled *burette*.

C

Calorie – the amount of heat required to raise the temperature of one gram of water from 14.5 °C to 15.5 °C; 1 calorie = 4.184 joules.

Calorimeter – a device used to measure the heat transfer between a system and its surroundings.

Canal ray – a stream of positively charged particles (cations) that moves toward the negative electrode in cathode ray tubes; observed to pass through canals in the negative electrode.

Capillary – a tube having a very small inside diameter.

Capillary action – the drawing of a liquid up the inside of a small-bore tube when adhesive forces exceed cohesive forces; the depression of the surface of the liquid when cohesive forces exceed the adhesive forces.

Catalyst – a chemical compound used to change the rate (to speed or slow it) of a regenerated reaction (i.e., not consumed) at the end of the reaction.

Catenation – the bonding of atoms of the same element into chains or rings (i.e., the ability of an element to bond with itself).

Cathode – the electrode at which reduction occurs; in a cathode ray tube, the negative electrode.

Cathodic protection – protection of a metal (making a cathode) against corrosion by attaching it to a sacrificial anode of a more easily oxidized metal.

Cathode ray tube – a closed glass tube containing gas under low pressure, with electrodes near the ends and a luminescent screen near the positive electrode; produces cathode rays when a high voltage is applied.

Cation – a positive ion; an atom or group of atoms that has lost one or more electrons.

Cell potential – the potential difference, E_{cell}, between oxidation and reduction half-cells under nonstandard conditions; the force in a galvanic cell pulls electrons through a reducing agent to an oxidizing agent.

Central atom – an atom in a molecule or polyatomic ion bonded to more than one other atom.

Chain reaction – a reaction that, once initiated, sustains itself and expands; a reaction in which reactive species, such as radicals, are produced in more than one step, allowing these reactive species to propagate the chain reaction.

Charles' Law – at constant pressure, the volume occupied by a definite mass of gas is directly proportional to its absolute temperature.

Chemical bonds – the attractive forces holding atoms together in elements or compounds.

Chemical change – when one or more new substances are formed.

Chemical equation – description of a chemical reaction by placing the formulas of the reactants on the left of an arrow and the formulas of the products on the right.

Chemical equilibrium – a state of dynamic balance in which the rates of forward and reverse reactions are equal; there is no net change in concentrations of reactants or products while a system is at equilibrium.

Chemical kinetics – studies rates and mechanisms of chemical reactions and factors they depend on.

Chemical periodicity – the variations in properties of elements with their position in the periodic table.

Chemical reaction – the change of one or more substances into another or multiple substances.

Cloud chamber – a device for observing the paths of speeding particles as vapor molecules condense on them to form fog-like tracks.

Cobalt chloride paper – water test; water changes the color from blue to pink.

Coefficient of expansion – the ratio of the change in the length or the volume of a body to the original length or volume for a unit change in temperature.

Cohesive forces – the forces of attraction among particles of a liquid.

Colligative properties – physical properties of solutions that depend upon the number but not the kind of solute particles present.

Collision theory – the theory of reaction rates that states that effective collisions between reactant molecules must occur for the reaction to occur.

Colloid – a heterogeneous mixture in which solute-like particles do not settle (e.g., many kinds of milk).

Combination reaction – two substances (elements or compounds) combine to form one compound.

Combustible – classification of liquid substances that burn based on flashpoints; any liquid having a flashpoint at or above 37.8 °C (100 °F) but below 93.3 °C (200 °F), except any mixture having components with flashpoints of 93.3 °C (200 °F) or higher, the total of which makes up 99% or more of the volume of the mixture.

Combustion (or *burning*) – an exothermic reaction between an oxidant and fuel with heat and often light.

Common ion effect – suppression of ionization of a weak electrolyte by the presence in the same solution of a strong electrolyte containing one of the same ions as the weak electrolyte.

Complex ions – ions resulting from coordinating covalent bonds between simple ions and other ions or molecules.

Composition stoichiometry – describes the quantitative (mass) relationships among elements in compounds.

Compound – a substance of two or more chemically bonded elements in fixed proportions; can be decomposed into constituent elements.

Compressed gas – a single or mixture of gases having (in a container) an absolute pressure exceeding 40 psi at 21.1 °C (70 °F).

Compression – an area in a longitudinal wave where the particles are closer and pushed in.

Concentration – the amount of solute per unit volume, the mass of solvent or solution.

Condensation – the phase change from gas to liquid.

Condensed phases – the liquid and solid phases; phases in which particles interact strongly.

Condensed states – the solid and liquid states.

Conduction – heat transfer between substances in direct contact with each other (i.e., must be touching); when particles of a hotter substance vibrate, these molecules bump into nearby particles and transfer some energy.

Conduction band – a vacant or partially filled band of energy levels just higher in energy than a filled band; a band within which, or into which, electrons must be promoted to allow electrical conduction to occur in a solid.

Conductor – a material that allows electric flow more freely.

Conjugate acid-base pair – in Brønsted-Lowry terms, reactant and product that differ by a proton, H^+.

Conformations – structures of a compound that differ by the extent of rotation about a single bond.

Continuous spectrum – contains wavelengths in a specified region of the electromagnetic spectrum.

Control rods – rods of materials such as cadmium or boron steel that act as neutron absorbers (not merely moderators), used in nuclear reactors to control neutron fluxes and therefore fission rates.

Conjugated double bonds – double bonds separated from each other by one single bond $-C=C-C=C-$

Contact process – the industrial process for sulfur trioxide (SO_3) and sulfuric acid (H_2SO_4) production from sulfur dioxide (SO_2).

Convection – the physical flow of matter when heat flows by energized molecules from one place to another through the movement of fluids. Transfer of heat through a liquid or a gas occurs when molecules of the liquid or gas move and carry the heat.

Coordinate covalent bond – a covalent bond with shared electrons furnished by the same species; a bond between a Lewis acid and a Lewis base.

Coordination compound or complex – a compound containing coordinate covalent bonds.

Coordination number – the number of donor atoms coordinated to a metal; the number of nearest neighbors of an atom or ion in describing crystals.

Coordination sphere – the metal ion and its coordinating ligands, but no uncoordinated counter-ions.

Corrosion – oxidation of metals (e.g., rusting) in the presence of air and moisture.

Coulomb – the SI unit of electrical charge; unit symbol – C.

Covalent bond – a force of attraction (chemical bond) formed by the sharing of electron pairs between two atoms.

Covalent compounds – compounds made of two or more nonmetal atoms bonded by sharing valence electrons.

Critical mass – the minimum mass of a particular fissionable nuclide in a given volume required to sustain a nuclear chain reaction.

Critical point – the combination of critical temperature and critical pressure of a substance.

Critical pressure – the pressure required to liquefy a gas (vapor) at its *Critical temperature.*

Critical temperature – the temperature above which a gas cannot be liquefied; the temperature above which a substance cannot exhibit distinct gas and liquid phases.

Crystal – a solid packed with ions, molecules, or atoms in an orderly fashion.

Crystal field stabilization energy – a measure of the net energy of stabilization gained by a metal ion's nonbonding d electrons due to complex formation.

Crystal field theory – bonding in transition metal complexes in which ligands and metal ions are treated as point charges; a purely ionic model; ligand point charges represent the crystal (electrical) field perturbing the metal's d orbitals containing nonbonding electrons.

Crystal lattice – a pattern of arrangement of particles in a crystal.

Crystal lattice energy – the amount of energy that holds a crystal together; the energy change when a mole of solid forms from its constituent molecules or ions (for ionic compounds) in their gaseous state (consistently negative).

Crystalline solid – a solid characterized by a regular, ordered arrangement of particles.

Curie (Ci) – the basic unit to describe the intensity of radioactivity in a sample of material; one curie equals 37 billion disintegrations per second or approximately the amount of radioactivity given off by 1 gram of radium.

Current – a flow of charged particles, such as electrons or ions, moving through an electrical space or conductor. It is measured as the net rate of flow of electric charge; the unit is an Ampere (A).

Cuvette – glassware used in spectroscopic experiments; usually made of plastic, glass, or quartz, and should be as clean and transparent as possible.

Cyclotron – a device for accelerating charged particles along a spiral path.

D

Dalton's Law (or the *law of partial pressures*) – the pressure exerted by a mixture of gases is the sum of the partial pressures of the individual gases.

Daughter nuclide – nuclide produced in nuclear decay.

Debye – the unit used to express dipole moments.

Degenerate – in orbitals, describes orbitals of the same energy.

Deionization – the removal of ions; in the case of water, mineral ions such as sodium, iron, and calcium.

Deliquescence – substances that absorb water from the atmosphere to form liquid solutions.

Delocalization – in reference to electrons, bonding electrons distributed among more than two atoms bonded; occurs in species that exhibit resonance.

Density – mass per unit volume; $D = M \times V$.

Deposition – settling particles within a solution; the direct solidification of vapor by cooling; see *sublimation*.

Derivative – a compound that can be imagined arising from a parent compound by replacing one atom with another atom or group of atoms; used extensively in organic chemistry to identify compounds.

Detergent – a soap-like emulsifier with sulfate, SO_3, or a phosphate group instead of carboxylate group.

Deuterium – an isotope of hydrogen whose atoms are twice as massive as ordinary hydrogen; deuterium atoms contain a proton and a neutron in the nucleus.

Dextrorotatory – refers to an optically active substance that rotates plane-polarized light clockwise, also known as *dextro*.

Diagonal similarities – chemical similarities in the Periodic Table of Elements of Period 2 to elements of Period 3 one group to the right, especially evident toward the left of the periodic table.

Diamagnetism – weak repulsion by a magnetic field.

Differential Scanning Calorimetry (DSC) – a technique for measuring temperature, direction, and magnitude of thermal transitions in a sample material by heating/cooling and comparing the amount of energy required to maintain its rate of temperature increase or decrease with an inert reference material under similar conditions.

Differential Thermal Analysis (DTA) – a technique for observing the temperature, direction, and magnitude of thermally induced transitions in a material by heating/cooling a sample and comparing its temperature with an inert reference material under similar conditions.

Differential thermometer – a thermometer used to measure minimal temperature changes accurately.

Dilution – the process of reducing the concentration of a solute in a solution, usually by mixing with more solvent.

Dimer – a molecule formed by combining two smaller (identical) molecules.

Dipole – electric or magnetic separation of charge; charge separation between two covalently bonded atoms.

Dipole-dipole interactions – attractive electrostatic forces between polar molecules (i.e., between molecules with permanent dipoles).

Dipole moment – the product of the distance separating opposite charges of equal magnitude; a measure of the polarity of a bond or molecule; a measured dipole refers to the dipole moment of an entire molecule.

Dispersing medium – the solvent-like phase in a colloid.

Dispersed phase – the solute-like species in a colloid.

Displacement reactions – reactions in which one element displaces another from a compound.

Disproportionation reactions – redox reactions in which the oxidizing agent and the reducing agent are the same species.

Dissociation – in an aqueous solution, the process by which a solid ionic compound separates into its ions.

Dissociation constant – equilibrium constant for dissociating a complex ion into a simple ion and coordinating species (ligands).

Dissolution or solvation – the spread of ions in a monosaccharide.

Distilland – the material in a distillation apparatus that is to be distilled.

Distillate – the material in a distillation apparatus collected in the receiver.

Distillation – separating a liquid mixture into its components based on differences in boiling points; the process in which components of a mixture are separated by boiling away the more volatile liquid; the vaporization of a liquid by heating and then the condensation of the vapor by cooling.

Domain – a cluster of atoms in a ferromagnetic substance, which align in the same direction in the presence of an external magnetic field.

Donor atom – a ligand atom whose electrons are shared with a Lewis acid.

***d*-orbitals** – beginning in the third energy level, a set of five degenerate orbitals per energy level, higher in energy than *s* and *p* orbitals of the same energy level.

Dosimeter – a small, calibrated electroscope worn by laboratory personnel to measure incident ionizing radiation or chemical exposure.

Double bond – covalent bond resulting from the sharing of four electrons (two pairs) between two atoms.

Double salt – a solid consisting of two co-crystallized salts.

Doublet – two peaks or bands of about equal intensity appearing close on a spectrogram.

Downs cell – an electrolytic cell for the commercial electrolysis of molten sodium chloride.

DP number – the degree of polymerization; the average number of monomer units per polymer unit.

Dry cells (voltaic cells) – ordinary batteries for appliances (e.g., flashlights, radios).

Dumas method – a method used to determine the molecular weights of volatile liquids.

Dynamic equilibrium – an equilibrium in which the processes occur continuously with no net change.

E

Earth metal – highly reactive elements in group IIA of the periodic table (includes beryllium, magnesium, calcium, strontium, barium, and radium); see *alkaline earth metal*.

Effective collisions – a collision between molecules resulting in a reaction, one in which the molecules collide with proper relative orientations and sufficient energy to react.

Effective molality – the sum of the molalities of solute particles in a solution.

Effective nuclear charge – the nuclear charge experienced by the outermost electrons of an atom; the actual nuclear charge minus the effects of shielding due to inner-shell electrons (e.g., a set of $d_{x^2-y^2}$ and d_{z^2} orbitals); those d orbitals within a set with lobes directed along the x, y, and z axes.

Electrical conductivity – the measure of how easily an electric current can flow through a substance.

Electric charge – a measured property (coulombs) that determines electromagnetic interaction.

Electrochemical cell – using a chemical reaction, an electromotive force is generated.

Electrochemistry – the study of chemical changes produced by electrical current and electricity production by chemical reactions.

Electrodes – surfaces upon which oxidation and reduction half-reactions occur in electrochemical cells; a conductor dips into an electrolyte and allows the electrons to flow to and from the electrolyte.

Electrode potentials – potentials, E, of half-reactions as reductions vs. standard hydrogen electrode.

Electrolysis – 1) occurs in electrolytic cells; chemical decomposition occurs by passing an electric current through a solution containing ions. 2) producing a chemical change using electricity; used to split up water into H and O_2.

Electrolyte – a substance (i.e., anions, cations) which, when dissolved in water, conducts electricity. An ionic solution that conducts a certain amount of current is categorized into two types: weak and strong.

Electrolytic cells – electrochemical cells in which electrical energy causes nonspontaneous redox reactions to occur (i.e., forced to occur by applying an outside source of electrical energy).

Electrolytic conduction – electrical current passes through ions in a solution or pure liquid.

Electromagnetic radiation – energy propagated using electric and magnetic fields that oscillate in directions perpendicular to the direction of travel of the energy; a type of wave that can go through vacuums as well as material; classified as a "self-propagating wave."

Electromagnetism – fields of an electric charge and electric properties that change how particles move and interact.

Electromotive force – a device that gains energy as electric charges pass through it.

Electromotive series – the relative order of tendencies for elements and their simple ions to act as oxidizing or reducing agents; also known as the "activity series."

Electron – a subatomic particle having a mass of 0.00054858 amu and a charge of -1.

Electron affinity – the energy absorbed in the process in which an electron is added to a neutral isolated gaseous atom to form a gaseous ion with a 1– charge; it has a negative value if energy is released.

Electron configuration – the specific distribution of electrons in atomic orbitals of atoms or ions.

Electron-deficient compounds – at least one atom (other than H) shares fewer than eight electrons.

Electron shells – an orbital around an atom's nucleus with a fixed number of electrons (usually 2 or 8).

Electronic transition – the transfer of an electron from one energy level to another.

Electronegativity – a measure of the relative tendency of an atom to attract electrons to itself when chemically combined with another atom.

Electronic geometry – the geometric arrangement of orbitals containing the shared and unshared electron pairs surrounding the central atom or polyatomic ion.

Electrophile – positively charged or electron-deficient.

Electrophoresis – separating ions by migration rate and direction of migration in an electric field.

Electroplating – a metal is covered with another metal layer using electricity; plating a metal onto a (cathodic) surface by electrolysis.

Element – a substance that cannot be decomposed into simpler substances by chemical means; defined by its *atomic number*. A substance that cannot be split into simpler substances by chemical means.

Eluant (or eluent) – the solvent used in the process of elution, as in liquid chromatography.

Eluate – a solvent (or mobile phase) which passes through a chromatographic column and removes the sample components from the stationary phase.

Emission spectrum – the emission of electromagnetic radiation by atoms (or other species) resulting from electronic transitions from higher to lower energy states.

Empirical formula – the simplest whole-number ratio of atoms of each element present in a compound; also known as the *simplest formula*.

Emulsifying agent – a substance that coats the particles of the dispersed phase and prevents coagulation of colloidal particles; an emulsifier.

Emulsion – colloidal suspension of a liquid in a liquid.

Endothermic – describes processes that absorb heat energy (H).

Endothermicity – the absorption of heat by a system as the process occurs.

Endpoint – the point at which an indicator changes color and titration stops.

Energy – a system's ability to do work.

Enthalpy (*H*) – the heat content of a specific amount of substance; $E = PV$.

Entropy (*S*) – a thermodynamic state or property that measures the degree of disorder (i.e., randomness) of a system; the amount of energy not available for work in a closed thermodynamic system (usually denoted by S).

Enzyme – a protein that acts as a catalyst in biological systems.

Equation of state – an expression that describes the behavior of matter in a given state; the van der Waals equation describes the behavior of the gaseous state.

Equilibrium or chemical equilibrium – a state of dynamic balance with the rates of forward and reverse reactions equal; the state of a system where neither forward nor reverse reaction is thermodynamically favored.

Equilibrium constant – a quantity characterizing equilibrium position for a reversible reaction; its magnitude is equal to mass action expression at equilibrium; equilibrium, "K," varies with temperature.

Equivalence point – the point when chemically equivalent amounts of reactants have reacted.

Equivalent weight – an oxidizing or reducing agent whose mass gains (oxidizing agents) or loses (reducing agents) 6.022×10^{23} electrons in a redox reaction.

Evaporation – vaporization of a liquid below its boiling point.

Evaporation rate – the rate at which a particular substance vaporizes (evaporate) compared to the rate of a known substance, such as ethyl ether, especially useful for health and fire-hazard considerations.

Excited state – any state other than the ground state of an atom or molecule; see *ground state*.

Exothermic – describes processes that release heat energy (*H*).

Exothermicity – the release of heat by a system as a process occurs.

Explosive – a chemical or compound that causes a sudden, almost instantaneous release of pressure, gas, heat, and light when subjected to sudden shock, pressure, high temperature, or applied potential.

Explosive limits – the range of concentrations over which a flammable vapor mixed with the proper ratios of air will ignite or explode if a source of ignition is provided.

Extensive property – a property that depends upon the amount of material in a sample.

Extrapolate – to estimate the value of a result outside the range of a series of known values; a technique used in the standard additions calibration procedure.

F

Faraday constant – a unit of electrical charge widely used in electrochemistry and equal to ~ 96,500 coulombs; represents 1 mole of electrons, or the Avogadro number of electrons: 6.022×10^{23} electrons.

Faraday's law of electrolysis – a two-part law that Michael Faraday published about electrolysis. 1. the mass of a substance altered at an electrode during electrolysis is directly proportional to the quantity of electricity transferred at that electrode. 2. the mass of an elemental material altered at an electrode is directly proportional to the element's equivalent weight; one equivalent weight of a substance is produced at each electrode during the passage of 96,487 coulombs of charge through an electrolytic cell.

Fast neutron – a neutron ejected at high kinetic energy in a nuclear reaction.

Ferromagnetism – the ability of a substance to become permanently magnetized by exposure to an external magnetic field.

Flashpoint – the temperature at which a liquid yields enough flammable vapor to ignite; various recognized industrial testing methods exist; therefore, the method used must be specified.

Fluorescence – absorption of high-energy radiation and subsequent emission of visible light.

First Law of Thermodynamics – the amount of energy in the universe is constant (i.e., energy is neither created nor destroyed in ordinary chemical reactions and physical changes); known as the Law of Conservation of Energy.

Fluids – substances that flow freely; gases and liquids.

Flux – a substance added to react with the charge or a product of its reduction; in metallurgy, it is usually added to lower the melting point.

Foam – colloidal suspension of a gas in a liquid.

Formal charge – a method of counting electrons in a covalently bonded molecule or ion; it counts bonding electrons as though they were equally shared between the two atoms.

Formula – a combination of symbols that indicates the chemical composition of a substance.

Formula unit – the smallest repeating unit of a substance; the molecule for nonionic substances.

Formula weight – the mass of one formula unit of a substance in atomic mass units.

Fossil fuels – formed from the remains of plants and animals that lived millions of years ago.

Fractional distillation – when a fractioning column is used in a distillation apparatus to separate the components of a liquid mixture with different boiling points.

Fractional precipitation – removal of some ions from a solution by precipitation while leaving other ions with similar properties in the solution.

Free energy change – the indicator of the spontaneity of a process at constant T and P (e.g., if ΔG is negative, the process is spontaneous).

Free radical – a highly reactive chemical species carrying no charge and having a single unpaired electron in an orbital.

Freezing – phase transition from liquid to solid.

Freezing point depression – the decrease in the freezing point of a solvent caused by the presence of a solute.

Frequency – the number of repeating points on a wave that passes a given observation point per unit time; the unit is 1 hertz = 1 cycle per 1 second.

Fuel – any substance that burns in oxygen to produce heat.

Fuel cells – a voltaic cell that converts the chemical energy of a fuel and an oxidizing agent directly into electrical energy continuously.

G

Gamma (γ) ray – a highly penetrating type of nuclear radiation similar to X-ray radiation, except that it comes from within the nucleus of an atom and has higher energy; energy-wise, very similar to cosmic rays, except that cosmic rays originate from outer space.

Galvanic cell – a battery made from an electrochemical cell with two metals connected by a salt bridge.

Galvanizing – placing a thin layer of zinc on a ferrous material to protect the underlying surface from corrosion.

Gangue – sand, rock, and other impurities surrounding the mineral of interest in an ore.

Gas – a state of matter in which the particles have no definite shape or volume, though they fill their container.

Gay-Lussac's Law – the expression used for each of the two relationships named after the French chemist Joseph Louis Gay-Lussac concerning the properties of gases; more usually applied to his law of combining volumes.

Geiger counter – a gas-filled tube that discharges electrically when ionizing radiation passes through it.

Gel – colloidal suspension of a solid dispersed in a liquid; a semi-rigid solid.

Gibbs (free) energy – the thermodynamic state function of a system that indicates the amount of energy available for the system to do useful work at constant T and P; a value that indicates the spontaneity of a reaction (usually denoted by *G*).

Graham's Law – the rates of effusion of gases are inversely proportional to the square roots of their molecular weights or densities.

Ground state – the lowest energy state or most stable state of an atom, molecule, or ion; see *excited state*.

Group – a vertical column in the periodic table; known as a family.

H

Haber process – a process for the catalyzed industrial production of ammonia from N_2 and H_2 at high temperature and pressure.

Half-cell – the compartment in which the oxidation or reduction half-reaction occurs in a voltaic cell.

Half-life – the time required for half of a reactant to be converted into product(s); the time required for half of a given sample to undergo radioactive decay.

Half-reaction – the oxidation or the reduction part of a redox reaction.

Halogens – group VIIA elements: F, Cl, Br, I; halogens are nonmetals.

Hard water – water high in dissolved minerals that makes it difficult to form lather with soap.

Heat – a form of energy that flows between two samples of matter because of temperature differences.

Heat capacity – the amount of heat required to raise the temperature of a mass one degree Celsius.

Heat of condensation – the amount of heat that must be removed from one gram of vapor at its condensation point to condense the vapor with no change in temperature.

Heat of crystallization – the amount of heat that must be removed from one gram of a liquid at its freezing point to freeze it with no change in temperature.

Heat of fusion – the amount of heat required to melt one gram of a solid at its melting point with no change in temperature; usually expressed in J/g; the molar heat of fusion is the amount of heat required to melt one mole of a solid at its melting point with no change in temperature and is usually expressed in kJ/mol.

Heat of solution – the amount of heat absorbed in forming a solution that contains one mole of the solute; the value is positive if heat is absorbed (endothermic) and negative if heat is released (exothermic).

Heat of vaporization – the amount of heat required to vaporize one gram of a liquid at its boiling point with no change in temperature; usually expressed in J/g; the molar heat of vaporization is the amount of heat required to vaporize one mole of liquid at its boiling point with no change in temperature and is usually expressed as ion kJ/mol.

Heisenberg uncertainty principle – states that it is impossible to determine the momentum and position of an electron simultaneously with absolute accuracy.

Henry's Law – the gas pressure above a solution is proportional to concentration of the gas in solution.

Hess' Law of heat summation – the enthalpy change for a reaction is the same whether it occurs in one step or a series of steps.

Heterogeneous catalyst – exists in a different phase (solid, liquid, or gas) from the reactants; a contact catalyst.

Heterogeneous equilibria – equilibria involving species in more than one phase.

Heterogeneous mixture – a mixture that does not have uniform composition and properties throughout.

Heteronuclear – consisting of different elements.

High spin complex – crystal field designation for an outer orbital complex; t_{2g} and e_g orbitals are singly occupied before pairing occurs.

Homogeneous catalyst – in the same phase (solid, liquid, or gas) as the reactants.

Homogeneous equilibria – when *reagents* and *products* are in the same phase (gases, liquids, or solids).

Homogeneous mixture – a mixture which has uniform composition and properties throughout.

Homologous series – compounds with each member differing from the next by a specific number and kind of atoms.

Homonuclear – consisting of only one element.

Hund's rule – single electrons must occupy orbitals of a given sublevel before pairing begins; see *Aufbau* (or *building up*) *principle*.

Hybridization – mixing atomic orbitals to form a new set of atomic orbitals with the same electron capacity and properties and energies intermediate between the original unhybridized orbitals.

Hydrate – a solid compound that contains a definite percentage of bound water.

Hydrate isomers – crystalline complexes that differ in whether water exists inside or outside the coordination sphere.

Hydration – the reaction of a substance with water.

Hydration energy – the energy change accompanying the hydration of a mole of gas and ions.

Hydride – a binary compound of hydrogen.

Hydrocarbons – compounds that contain only carbon and hydrogen.

Hydrogen bond – a relatively strong dipole-dipole interaction (but still considerably weaker than the covalent or ionic bonds) between molecules containing hydrogen directly bonded to a small, highly electronegative atom, such as N, O or F.

Hydrogenation – the reaction in which hydrogen adds across a double or triple bond.

Hydrogen-oxygen fuel cell – hydrogen is the fuel (reducing agent), and oxygen is the oxidizing agent.

Hydrolysis – the reaction of a substance with water or its ions.

Hydrolysis constant – an equilibrium constant for a hydrolysis reaction.

Hydrometer – a device used to measure the densities of liquids and solutions.

Hydrophilic colloids – colloidal particles that repel water molecules.

I

Ideal gas – a hypothetical gas that obeys the postulates of the Kinetic Molecular Theory.

Ideal gas law – the product of pressure and the volume of an ideal gas is directly proportional to the number of moles of the gas and the absolute temperature. $PV = nRT$

Ideal solution – obeys Raoult's Law strictly.

Immiscible liquids – do not mix to form a solution (e.g., oil and water).

Indicators – for acid-base titrations, organic compounds that exhibit different colors in solutions of different acidities, determine the point at which the reaction between two solutes is complete.

Inert pair effect – characteristic of the post-transition minerals; the tendency of the electrons in the outermost atomic *s* orbital to remain un-ionized or unshared in compounds of post-transition metals.

Inhibitory catalyst – an inhibitor; a catalyst that decreases the rate of reaction.

Inner orbital complex – valence bond designation for a complex in which the metal ion utilizes d orbitals for one shell inside the outermost occupied shell in its hybridization.

Inorganic chemistry – a part of chemistry concerned with inorganic (non-carbon-based) compounds.

Insulator – a material that resists the flow of electric current or heat transfer; it does not allow heat to flow easily.

Insoluble compound – a substance that will not dissolve in a solvent, even after mixing.

Integrated rate equation – an expression giving the concentration of a reactant remaining after a specified time; it has a different mathematical form for different orders of reactants.

Intermolecular forces – forces between individual particles (atoms, molecules, ions) of a substance.

Ion – a molecule that has gained or lost electrons; an atom or a group of atoms carries an electric charge (Na^+).

Ion product for water – equilibrium constant for water ionization; $Kw = [H_3O^+] \cdot [^-OH] = 1.00 \times 10^{-14}$ at 25 °C.

Ionic bond – the electrostatic attraction between oppositely charged ions, resulting from a transfer of electrons.

Ionic bonding – chemical bonding resulting from transferring electrons from one atom or group.

Ionic compounds – compounds containing predominantly ionic bonding.

Ionic geometry – arrangement of atoms (not lone pairs of electrons) about the central atom of a polyatomic ion.

Ionization – the breaking up of a compound into separate ions; in an aqueous solution, the process by which a molecular compound reacts with water and forms ions.

Ionization constant – equilibrium constant for the ionization of a weak electrolyte.

Ionization energy – the minimum amount of energy required to remove the most loosely held electron of an isolated gaseous atom or ion.

Ion exchange – a method of removing hardness from water, it replaces the positive ions that cause the hardness with H^+ ions.

Ionization isomers – result from the interchange of ions inside and outside the coordination sphere.

Isoelectric – having the same electronic configurations.

Isomers – different substances with the same formula.

Isomorphous – refers to crystals having the same atomic arrangement.

Isotopes – two or more forms of atoms of the same element with different masses; atoms containing the same number of protons but different numbers of neutrons.

IUPAC – acronym for "International Union of Pure and Applied Chemistry."

J

Joule (J) – a unit of energy in the SI system; one joule is $1 \text{ kg} \cdot \text{m}^2/\text{s}^2$, which is 0.2390 calories.

K

K capture – absorption of a K shell (n = 1) electron by a proton as it is converted to a neutron.

Kelvin – a unit of measure for temperature based upon an absolute scale.

Kinetics – a subfield of chemistry specializing in reaction rates.

Kinetic energy (*KE*) – energy that matter processes through its motion.

Kinetic Molecular Theory – a theory that attempts to explain macroscopic observations on gases in microscopic or molecular terms.

L

Lanthanides – elements 57 (lanthanum) through 71 (lutetium); grouped because of their similar behavior in chemical reactions.

Lanthanide contraction – a decrease in the radii of the elements following the lanthanides compared to what would be expected if there were no *f*-transition metals.

Latent heat – the energy absorbed or released when a substance changes state without changing temperature.

Lattice – a unique arrangement of atoms or molecules in a crystalline liquid or solid.

Law of combining volumes (Gay-Lussac's Law) – at constant temperature and pressure, the volumes of reacting gases (and any gaseous products) can be expressed as ratios of small whole numbers.

Law of conservation of energy – energy cannot be created or destroyed; it can only change form.

Law of conservation of matter – there is no detectable change in the quantity of matter during an ordinary chemical reaction.

Law of conservation of matter and energy – the amount of matter and energy in the universe is fixed.

Law of definite proportions (law of constant composition) – different samples of a pure compound contain the same elements in the same proportions by mass.

Law of partial pressures (or *Dalton's Law*) – the pressure exerted by a mixture of gases is the sum of the partial pressures of the individual gases.

Laws of thermodynamics – physical laws which define quantities of thermodynamic systems describe how they behave and (by extension) set certain limitations, such as perpetual motion.

Lead storage battery – a secondary voltaic cell used in most automobiles.

Leclanche cell – a common type of *dry cell*.

Le Chatelier's principle – states that a system at equilibrium, or striving to attain equilibrium, responds in such a way as to counteract any stress placed upon it; if stress (change of conditions) is applied to a system at equilibrium, the system will shift in the direction that reduces stress.

Leveling effect – acids stronger than the acid characteristic of the solvent react with the solvent to produce that acid; a similar statement applies to bases. The strongest acid (base) that can exist in a given solvent is the acid (base) characteristic of the solvent.

Levorotatory (or *levo*) – an optically active substance rotates plane-polarized light counterclockwise.

Lewis acid – any species that can accept a share in an electron pair.

Lewis base – any species that can make available a pair of electrons.

Lewis dot formula (electron dot formula) – representation of a molecule, ion, or formula unit by showing atomic symbols and only outer shell electrons.

Ligand – a Lewis base in a coordination compound.

Light – that portion of the electromagnetic spectrum visible to the naked eye; known as "visible light."

Limiting reactant – a substance that stoichiometrically limits the number of products formed.

Linear accelerator – a device used for accelerating charged particles along a straight line path.

Line spectrum – an atomic emission or absorption spectrum.

Linkage isomers – a particular ligand bonds to a metal ion through different donor atoms.

Liquid – a state of matter which takes the shape of its container.

Liquid aerosol – colloidal suspension of a liquid in a gas.

London dispersion forces – very weak and very short-range attractive forces between short-lived temporary (induced) dipoles; known as "dispersion forces."

Lone pair – a pair of electrons residing on one atom and not shared by other atoms; an unshared pair.

Low spin complex – crystal field designation for an inner orbital complex; contains electrons paired t_{2g} orbitals before e_g orbitals are occupied in octahedral complexes.

Lubricant – a substance capable of reducing friction (i.e., a force that opposes the direction of motion).

M

Magnetic field – a space around a magnet where magnetism can be detected.

Magnetic quantum number – quantum mechanical solution to a wave equation designating the orbital within a given set (s, p, d, f) in which an electron resides.

Manometer – a two-armed barometer.

Mass (m) – a measure of the amount of matter in an object; mass is usually measured in grams or kilograms.

Mass action expression – for a reversible reaction, aA + bB cC + dD; the product of the concentrations of the products (species on the right), each raised to the power corresponding to its coefficient in the balanced chemical equation, divided by the product of the concentrations of reactants (species on the left), each raised to the power corresponding to its coefficient in the balanced equation; at equilibrium the mass action expression is equal to K.

Mass deficiency – the amount of matter converted into energy when an atom forms from constituent particles.

Mass number – the sum of the numbers of protons and neutrons in an atom; an integer.

Mass spectrometer – an instrument that measures the charge-to-mass ratio of charged particles.

Matter – anything that has mass and occupies space.

Mechanism – the sequence of steps by which reactants are converted into products.

Melting point – the temperature at which liquid and solid coexist in equilibrium.

Meniscus – the shape assumed by the surface of a liquid in a cylindrical container.

Melting – the phase change from a solid to a liquid.

Metal – a chemical element that is a good conductor of electricity and heat and forms cations and ionic bonds with nonmetals; elements below and to the left of the stepwise division (metalloids) in the upper right corner of the periodic table; about 80% of known elements are metals.

Metallic bonding – bonding within metals due to the electrical attraction of positively charged metal ions for mobile electrons that belong to the crystal.

Metallic conduction – conduction of electrical current through a metal or along a metallic surface.

Metalloid – a substance with the properties of metals and nonmetals (B, Al, Si, Ge, As, Sb, Te, Po and At).

Metathesis reactions – reactions in which two compounds react to form two new compounds, with no changes in oxidation number; reactions in which the ions of two compounds exchange partners.

Method of initial rates – method of determining the rate-law expression by carrying out a reaction with different initial concentrations and analyzing the resultant changes in initial rates.

Methylene blue – a heterocyclic aromatic chemical compound with the molecular formula $C_{16}H_{18}N_3SCl$.

Miscible liquids – mix to form a solution (e.g., alcohol and water).

Miscibility – the ability of one liquid to mix with (dissolve in) another liquid.

Mixture – two or more different substances mingled together but not chemically combined. A sample of matter composed of two or more substances, each of which retains its identity and properties.

Moderator – a substance (e.g., deuterium, oxygen, paraffin) that slows fast neutrons upon collision.

Molality – a concentration expressed as the number of moles of solute per kilogram of solvent.

Molarity – the number of moles of solute per liter of solution.

Molar solubility – the number of moles of a solute that dissolves to produce a liter of a saturated solution.

Mole – a measurement of an amount of substance; a single mole contains approximately 6.022×10^{23} units or entities; abbreviated mol.

Molecule – a chemically bonded number of electrically neutral atoms.

Molecular equation – a chemical reaction in which formulas are written as if substances existed as molecules; only complete formulas are used.

Molecular formula – indicates the number of atoms present in a molecule of a molecular substance.

Molecular geometry – the arrangement of atoms (not lone pairs of electrons) around a central atom of a molecule or polyatomic ion.

Molecular orbital (*MO*) – resulting from the overlap and mixing of atomic orbitals on different atoms (i.e., a region where an electron can be found in a molecule, as opposed to an atom); an MO belongs to the molecule.

Molecular orbital theory – a theory of chemical bonding based upon postulated molecular orbitals.

Molecular weight – the mass of one molecule of a nonionic substance in atomic mass units.

Molecule – the smallest particle of a compound capable of stable, independent existence.

Mole fraction – the number of moles of a component in a mixture divided by the number of moles in the mixture.

Monoprotic acid – can form only one hydronium ion per molecule; may be strong or weak.

Mother nuclide – nuclide that undergoes nuclear decay.

N

Native state – refers to the occurrence of an element in an uncombined or free state in nature.

Natural radioactivity – spontaneous decomposition of an atom.

Neat – conditions with a liquid reagent or gas performed with no added solvent or co-solvent.

Nernst equation – corrects standard electrode potentials for nonstandard conditions.

Net ionic equation – results from canceling spectator ions and eliminating brackets from a total ionic equation.

Neutralization – the reaction of an acid with a base to form a salt and water; usually, the reaction of hydrogen ions with hydroxide ions to form water molecules.

Neutrino – a particle that travels at speeds close to the speed of light; created due to radioactive decay.

Neutron – a neutral unit or subatomic particle with no net charge and a mass of 1.0087 amu.

Nickel-cadmium cell (NiCd battery) – a dry cell where the anode is Cd, the cathode is NiO_2, and the electrolyte is basic.

Nitrogen cycle – the complex series of reactions by which nitrogen is slowly but continually recycled in the atmosphere, lithosphere, and hydrosphere.

Noble gases – elements of the periodic Group 0; He, Ne, Ar, Kr, Xe, Rn; known as "rare gases;" formerly called "inert gases."

Nodal plane – a region in which the probability of finding an electron is zero.

Nonbonding orbital – a molecular orbital derived only from an atomic orbital of one atom; it lends neither stability nor instability to a molecule or ion when populated with electrons.

Nonelectrolyte – a substance whose aqueous solutions do not conduct electricity.

Nonmetal – an element that is not metallic.

Nonpolar bond – a covalent bond in which electron density is symmetrically distributed.

Nuclear – of or about the atomic nucleus.

Nuclear binding energy – the energy equivalent of the mass deficiency; the energy released in forming an atom from the subatomic particles.

Nuclear fission – when a heavy nucleus splits into nuclei of intermediate masses, and protons are emitted.

Nuclear magnetic resonance spectroscopy – a technique that exploits the magnetic properties of specific nuclei; helps identify unknown compounds.

Nuclear reaction – involves a change in the composition of a nucleus and can emit or absorb a tremendous amount of energy.

Nuclear reactor – a system in which controlled nuclear fission reactions generate heat energy on a large scale, subsequently converted into electrical energy.

Nucleons – particles comprising the nucleus: protons and neutrons.

Nucleus – the tiny and dense, positively charged center of an atom containing protons and neutrons, as well as other subatomic particles; the net charge is positive.

Nuclides – refers to different atomic forms of elements; in contrast to isotopes, which refer only to different atomic forms of a single element.

Nuclide symbol – designation for an atom A/Z E, in which E is the symbol of an element, Z is its atomic number, and A is its mass number.

Number density – a measure of the concentration of countable objects (e.g., atoms, molecules) in space; the number per volume.

O

Octahedral – molecules and polyatomic ions with one atom in the center and six atoms at the corners of an octahedron.

Octane number – a number that indicates how smoothly a gasoline burns.

Octet rule – during bonding, atoms tend to reach an electron arrangement with eight electrons in the outermost shell. Many representative elements attain at least a share of eight electrons in their valence shells when they form molecular or ionic compounds; there are some limitations.

Open sextet – species with only six electrons in the highest energy level of the central element (many Lewis acids).

Orbital – may refer to an atomic orbital or a molecular orbital.

Organic chemistry – the chemistry of substances that contain carbon-hydrogen bonds.

Organic compound – substances that contain carbon.

Osmosis – when solvent molecules pass through a semi-permeable membrane from a dilute solution into a more concentrated solution.

Osmotic pressure – the hydrostatic pressure produced on the surface of a semi-permeable membrane.

Outer orbital complex – valence bond designation for a complex in which the metal ion utilizes d orbitals in the outermost (occupied) shell in hybridization.

Overlap – the interaction of orbitals on different atoms in the same region of space.

Oxidation – the addition of oxygen or the loss of electrons. An algebraic increase in the oxidation number may correspond to a loss of electrons.

Oxidation numbers – quantitative values used as mechanical aids in writing formulas and balancing equations; for single-atom ions, they correspond to the charge on the ion; more electronegative atoms are assigned negative oxidation numbers, known as *oxidation states*.

Oxidation-reduction reactions – reactions in which oxidation and reduction occur; known as *redox reactions*.

Oxide – a binary compound of oxygen.

Oxidizing agent – the substance that oxidizes another substance and is reduced.

P

Pairing – a favorable interaction of two electrons with opposite m values in the same orbital.

Pairing energy – the energy required to pair two electrons in the same orbital.

Paramagnetism – attraction toward a magnetic field, stronger than diamagnetism but still weak compared to ferromagnetism.

Partial pressure – the force exerted by one gas in a mixture of gases.

Particulate matter – fine, divided solid particles suspended in polluted air.

Pauli exclusion principle – no electrons in the same atom may have identical sets of four quantum numbers.

Percentage ionization – the percentage of the weak electrolyte that will ionize in a solution of given concentration.

Percent by mass – 100% times the actual yield divided by the theoretical yield.

Percent composition – the mass percent of each element in a compound.

Percent purity – the percent of a specified compound or element in an impure sample.

Period – the elements in a horizontal row of the periodic table.

Periodicity – regular periodic variations of properties of elements with their atomic number (and position in the periodic table).

Periodic Law – the properties of the elements are periodic functions of their atomic numbers.

Periodic table – an arrangement of elements by increasing atomic numbers, emphasizing periodicity.

Peroxide – a compound with oxygen in –1 oxidation state; metal peroxides contain the peroxide ion, O_2^{2-}.

pH – the measure of acidity (or basicity) of a solution; negative logarithm of the concentration (mol/L) of the H_3O^+ [H^+] ion; scale is commonly used over a range 0 to 14.

Phase diagram – shows the equilibrium temperature-pressure relationships for different phases of a substance.

pH scale – a range from 0 to 14. If the pH of a solution is 7, it is neutral; if the pH of a solution is less than 7, it is acidic; if the pH of a solution is greater than 7, it is basic.

Permanent hardness – hardness (relative to lathering soap) in water that cannot be removed by boiling; caused by calcium sulfate.

Photoelectric effect – emission of an electron from the surface of a metal caused by impinging electromagnetic radiation of specific minimum energy; the current increases with increasing radiation intensity.

Photon – a carrier of electromagnetic radiation of all wavelengths, such as gamma rays and radio waves; known as a *quantum of light*.

Physical change – when a substance changes from one physical state to another, but no substances with different compositions are formed; physical change may involve a phase change (e.g., melting, freezing) or another physical change, such as crushing a crystal or separating one volume of liquid into different containers; it does not produce a new substance.

Plasma – a physical state of matter that exists at extremely high temperatures in which molecules are dissociated, and most atoms are ionized.

Polar bond – a covalent bond with an unsymmetrical distribution of electron density.

Polarimeter – a device used to measure optical activity.

Polarization – the buildup of a product of oxidation or reduction of an electrode, preventing further reaction.

Polydentate – refers to ligands with more than one donor atom.

Polyene – a compound that contains more than one double bond per molecule.

Polymerization – the combination of many small molecules to form large molecules.

Polymer – a large molecule consisting of chains or rings of linked monomer units, usually characterized by high melting and boiling points.

Polymorphous – refers to substances that can crystallize in more than one crystalline arrangement.

Polyprotic acid – forms two or more H_3O+ ions per molecule; often, at least one ionization step is weak.

Positron – a nuclear particle with the mass of an electron but opposite charge (positive).

Potential difference (or *voltage*) – the force that moves the electrons around circuit; unit is Volt (V).

Potential energy (*PE*) – energy stored in a body or a system due to its position in a force field or configuration.

Power – the rate at which energy is converted from one form to another; the unit is Watts (W). Power = voltage × current (P = VI).

Precipitate – an insoluble solid formed by mixing in solution the constituent ions of a slightly soluble solution.

Precision – how close the results of multiple experimental trials are; see *accuracy*.

Pressure – force per unit area; unit is Pascal (Pa).

Primary standard – a known high degree of purity substance that undergoes one invariable reaction with the other reactant of interest.

Primary voltaic cells – voltaic cells that cannot be recharged; no further chemical reaction is possible once the reactants are consumed.

Products – chemicals produced (from reactants) in a chemical reaction.

Proton – a subatomic particle having a mass of 1.0073 amu and a charge of +1, found atom's nucleus.

Protonation – the addition of a proton (H^+) to an atom, molecule, or ion.

Pseudobinaryionic compounds – contain more than two elements but are named like binary compounds.

Q

Quanta – the minimum amount of energy emitted by radiation.

Quantum mechanics – the study of how atoms, molecules, subatomic particles, etc., behave and are structured; a mathematical method of treating particles based on quantum theory, which assumes that energy (of small particles) is not infinitely divisible.

Quantum numbers – numbers that describe the energies of electrons in atoms; derived from quantum mechanical treatment.

Quarks – elementary particles and a fundamental constituent of matter, combining to form hadrons (i.e., protons and neutrons).

R

Radiation – 1) heat transfer through invisible rays, which travel outwards from the hot object without a medium. 2) high-energy particles or rays emitted during the nuclear decay processes.

Radical – an atom or group of atoms that contains one or more unpaired electrons; usually a very reactive species.

Radioactive dating – a method of dating ancient objects by determining the ratio of mother and daughter nuclides present in an object and relating the ratio to the object's age via half-life calculations.

Radioactive tracer – a small amount of radioisotope replacing a nonradioactive isotope of the element in a compound whose path (e.g., in the body) or whose decomposition products are monitored by detection of radioactivity; known as a "radioactive label."

Radioactivity – the spontaneous disintegration of atomic nuclei.

Raoult's Law – the vapor pressure of a solvent in an ideal solution decreases as its mole fraction decreases.

Rate-determining step – the slowest step in a mechanism; determines the overall reaction rate.

Rate-law expression – an equation relating the reaction rate to the concentrations of the reactants and the specific rate of the reaction.

Rate of reaction – the change in the concentration of a reactant or product per unit time.

Reactants – substances consumed in a chemical reaction; react together in a chemical reaction.

Reaction quotient – the mass action expression under any set of conditions (not necessarily equilibrium); its magnitude relative to K determines the direction in which the reaction must occur to establish equilibrium.

Reaction ratio – the relative amounts of reactants and products involved in a reaction; may be the ratio of moles, millimoles, or masses.

Reaction stoichiometry – describes the quantitative relationships among substances participating in chemical reactions.

Reactivity series (or activity series) – an empirical, calculated, and structurally analytical progression of a series of metals, arranged by "reactivity" from highest to lowest; used to summarize information about the reactions of metals with acids and water, double displacement reactions, and the extraction of metals from ores.

Reagent – a substance (or compound) added to a system to cause a chemical reaction or to visualize if a reaction occurs; the terms reactant and reagent are often used interchangeably; however, a *reactant* is more specifically a substance consumed during a chemical reaction.

Reducing agent – a substance that reduces another substance and is itself oxidized.

Reduction – the removal of oxygen or the gaining of electrons.

Resonance – the concept in which two or more equivalent dot formulas for the same arrangement of atoms (resonance structures) are necessary to describe the bonding in a molecule or ion.

Reverse osmosis – forcing solvent molecules to flow through a semi-permeable membrane from a concentrated solution into a dilute solution by applying greater hydrostatic pressure on the concentrated side than the osmotic pressure opposing it.

Reversible reaction – processes that do not go to completion and occur in the forward and reverse direction.

S

Saline solution – a general term for NaCl (i.e., sodium chloride) in water.

Salt – when a metal replaces the hydrogen of an acid.

Salts – ionic compounds composed of anions and cations.

Salt bridge – a U-shaped tube containing an electrolyte, connects the two half-cells of a voltaic cell.

Saturated solution –no more solute will dissolve at that temperature.

***s*-block elements** – group 1 and 2 elements (alkali and alkaline metals), including hydrogen and helium.

Schrödinger equation – quantum state equation representing the behavior of an electron around an atom; describes the wave function of a physical system evolving.

Second Law of Thermodynamics – the universe tends toward a state of greater disorder in spontaneous processes.

Secondary standard – a solution that has been titrated against a primary standard; a standard solution.

Secondary voltaic cells – voltaic cells that can be recharged; original reactants can be regenerated by reversing the direction of the current flow.

Semiconductor – a substance that does not conduct electricity at low temperatures but will do so at higher temperatures.

Semi-permeable membrane – a thin partition between two solutions through which specific molecules can pass but others cannot.

Shielding effect – electrons in filled sets of s, p orbitals between the nucleus and outer shell electrons shield the outer shell electrons somewhat from the effect of protons in the nucleus; known as the "screening effect."

Sigma (σ) bonds – bonds resulting from the head-on overlap of atomic orbitals. The region of electron sharing is along and (cylindrically) symmetrical to the imaginary line connecting the bonded atoms.

Sigma orbital – molecular orbital resulting from the head-on overlap of two atomic orbitals.

Single bond – covalent bond resulting from the sharing of two electrons (one pair) between two atoms.

Sol – a suspension of solid particles in a liquid; artificial examples include sol-gels.

Solid – one of the states of matter, where the molecules are packed closely, resistance to movement/deformation, and volume change.

Solubility product constant – equilibrium constant for the dissolution of a slightly soluble compound.

Solubility product principle – the solubility product constant expression for a slightly soluble compound is the product of the concentrations of the constituent ions; each raised to the power that corresponds to the number of ions in one formula unit.

Solute – the dispersed (i.e., dissolved) phase of a solution; the solution is mixed into the solvent (e.g., NaCl in saline water).

Solution – a homogeneous mixture of multiple substances; comprised of solutes and solvents; a mixture of a solute (usually a solid) and a solvent (usually a liquid).

Solvation – the process by which solvent molecules surround and interact with solute ions or molecules.

Solvent – the dispersing medium of a solution (e.g., H_2O in saline water).

Solvolysis – the reaction of a substance with the solvent in which it is dissolved.

s-orbital – a spherically symmetrical atomic orbital; one per energy level.

Specific gravity – the ratio of the density of a substance to the density of water.

Specific heat – the amount of heat required to raise the temperature of one gram of substance 1 °C.

Specific rate constant – an experimentally determined (proportionality) constant; different for different reactions, and which changes only with temperature; k in the rate-law expression: Rate = k [A] × [B].

Spectator ions – ions in a solution that do not participate in a chemical reaction.

Spectral line – any of several lines corresponding to definite wavelengths of an atomic emission or absorption spectrum; marks the energy difference between two energy levels.

Spectrochemical series – arrangement of ligands in order of increasing ligand field strength.

Spectroscopy – the study of radiation and matter, such as X-ray absorption and emission spectroscopy.

Spectrum – display of component wavelengths (colors) of electromagnetic radiation.

Speed of light – the speed at which radiation travels through a vacuum (299,792,458 m/sec).

Square planar – describes molecules and polyatomic ions with one atom in the center and four atoms at the corners of a square.

Square planar complex – relationship with metal in the center of a square plane, with ligand donor atoms at each of the four corners.

Standard conditions for temperature and pressure (STP) – used to compare experimental results (25 °C and 100.000 kPa).

Standard electrodes – half-cells in which the oxidized and reduced forms of a species are present at the unit activity (1.0 M solutions of dissolved ions, 1.0 atm partial pressure of gases, pure solids, and liquids).

Standard electrode potential – by convention, the potential (Eo) of a half-reaction as a reduction relative to the standard hydrogen electrode when species are present at unit activity.

Standard entropy – the absolute entropy of a substance in its standard state at 298 K.

Standard molar enthalpy of formation – the amount of heat absorbed in forming one mole of a substance in a specified state from its elements in their standard states.

Standard molar volume – space occupied by 1 mole of ideal gas under standard conditions; 22.4 liters.

Standard reaction – a process where the numbers of moles of reactants in the balanced equation, in their standard states, are entirely converted to the numbers of moles of products in the balanced equation, at their standard state.

State of matter – a homogeneous, macroscopic phase (e.g., gas, plasma, liquid, solid) in increasing concentration.

Stoichiometry – quantitative relationships of elements and compounds undergoing chemical changes.

Strong electrolyte – a substance that conducts electricity well in a dilute aqueous solution.

Strong field ligand – a ligand that exerts a strong crystal or ligand electrical field and generally forms low-spin complexes with metal ions when possible.

Structural isomers – compounds that contain the same number and kinds of atoms, but with different geometries.

Subatomic particles – comprise an atom (e.g., protons, neutrons, electrons).

Sublimation – the direct vaporization of a solid by heating without passing through the liquid state; a phase transition from solid to gas.

Substance – any matter, specimens with the same chemical composition and physical properties.

Substitution reaction – a reaction in which another atom or group of atoms replaces an atom or a group of atoms.

Supercooled liquids – liquids that, when cooled, apparently solidify but continue to flow very slowly under the influence of gravity.

Supercritical fluid – a substance at a temperature above its critical temperature.

Supersaturated solution – contains a higher than saturation concentration of solute; slight disturbance or seeding causes crystallization of excess solute.

Suspension – a heterogeneous mixture in which solute-like particles settle out of the solvent-like phase sometime after their introduction. A mixture of a liquid and a finely divided insoluble solid.

T

Talc – a mineral representing the Mohs Scale, composed of hydrated magnesium silicate with the chemical formula $H_2Mg_3(SiO_3)_4$ or $Mg_3Si_4O_{10}(OH)_2$.

Temperature – a measure of heat intensity (i.e., hotness or coldness of a sample); a measure of kinetic energy of an object. Units are measured in degrees, and scales include Celsius, Fahrenheit, and Kelvin.

Temporary hardness – hardness in water, removed by boiling; caused by calcium hydrogen carbonate.

Ternary acid – a ternary compound containing H, O and another element, often a nonmetal.

Ternary compound – a compound consisting of three elements; may be ionic or covalent.

Tetrahedral – a term used to describe molecules and polyatomic ions with one atom in the center and four atoms at the corners of a tetrahedron.

Theoretical yield – the maximum amount of a specified product that could be obtained from specified amounts of reactants, assuming complete consumption of the limiting reactant according to only one reaction and complete recovery of the product; see *actual yield*.

Theory – a model describing the nature of a phenomenon.

Thermal conductivity – a property of a material to conduct heat (often noted as k).

Thermal cracking – decomposition by heating a substance in the presence of a catalyst and without air.

Thermochemistry – the study of absorption/release of heat within a chemical reaction; studies heat energy associated with chemical reactions and physical transformations.

Thermodynamics – studying the effects of changing temperature, volume, or pressure (or work, heat, and energy) on a macroscopic scale.

Thermodynamic stability – when a system is in its lowest energy state with its environment (equilibrium).

Thermometer – a device that measures the average energy of a system.

Thermonuclear energy – energy from nuclear fusion reactions.

Third Law of Thermodynamics – entropy of a pure crystalline substance at absolute zero equals zero.

Titration – a procedure in which one solution is added to another solution until the chemical reaction between the two solutions is complete; the concentration of one solution is known, and that of the other is unknown. The process of adding one solution to a measured amount of another to find out exactly how much of each is required to react.

Torr – a unit to measure pressure; 1 Torr is equivalent to 133.322 Pa or 1.3158×10^{-3} atm.

Total ionic equation – the expression for a chemical reaction written to show the predominant form of species in aqueous solution or contact with water.

Transition elements (metals) – B Group elements except IIB in the periodic table; sometimes called transition elements, elements with incomplete *d* sub-shells; the *d*-block elements.

Transition state theory –reactants pass through high-energy transition states before forming products.

Transuranic element – an atomic number greater than 92; none of the transuranic elements are stable.

Triple bond – the sharing of three pairs of electrons within a covalent bond (e.g., N_2).

Triple point – where the temperature and pressure of three phases are the same; water has a unique phase diagram.

Tyndall effect – results from light scattering by colloidal particles (a mixture where one substance is dispersed evenly throughout another) or by suspended particles.

U

Uncertainty –any measurement that involves estimating any amount that cannot be precisely reproduced.

Uncertainty principle – knowing the location of a particle makes the momentum uncertain, while knowing the momentum of a particle makes the location uncertain.

Unit cell – the smallest repeating unit of a lattice.

Unit factor – statements used in converting between units.

Universal (or ideal) gas constant – proportionality constant in the ideal gas law (0.08206 L·atm/(K·mol)).

UN number – a four-digit code used to note hazardous and flammable substances.

Unsaturated hydrocarbons – hydrocarbons that contain double or triple carbon-carbon bonds.

V

Valence bond theory – proposes that covalent bonds are formed when atomic orbitals on different atoms overlap and the electrons are shared.

Valence electrons – outermost electrons of atoms; usually those involved in bonding.

Valence shell electron pair repulsion theory (VSEPR) – assumes electron pairs are arranged around the central element of a molecule or polyatomic ion with maximum separation (and minimum repulsion) among regions of high electron density.

Valency – the number of electrons an atom wants to gain, lose, or share to have a full outer shell.

Van der Waals' equation – a quantitative relationship of a state extending the ideal gas law to real gases by including two empirically determined parameters, which are specific for different gases.

Van der Waals force – one of the forces (attraction/repulsion) between molecules.

Van't Hoff factor – the ratio of moles of particles in solution to moles of solute dissolved.

Vapor – when a substance is below the critical temperature in the gas phase.

Vaporization – the phase change from liquid to gas.

Vapor pressure – the particle pressure of vapor at the surface of its parent liquid.

Viscosity – the resistance of a liquid to flow (e.g., oil has a higher viscosity than water).

Volt – one joule of work per coulomb; the unit of electrical potential transferred.

Voltage – the potential difference between two electrodes; a measure of the chemical potential for a redox reaction.

Voltaic cells – electrochemical cells in which spontaneous chemical reactions produce electricity; known as *galvanic cells*.

Voltmeter – an instrument that measures the cell potential.

Volumetric analysis – measuring the volume of a solution (of known concentration) to determine the substance's concentration within the solution; see *titration*.

W

Water equivalent – the amount of water absorbing the same heat as the calorimeter per degree of temperature increase.

Weak electrolyte – a substance that conducts electricity poorly in a dilute aqueous solution.

Weak field ligand – a ligand that exerts a weak crystal or ligand field and generally forms high-spin complexes with metals.

X

X-ray – electromagnetic radiation between gamma and UV rays.

X-ray diffraction – a method for establishing structures of crystalline solids using single-wavelength X-rays and studying the diffraction pattern.

X-ray photoelectron spectroscopy – a spectroscopic technique used to measure the composition of a material.

Y

Yield – the amount of product produced during a chemical reaction.

Z

Zone melting – remove impurities from an element by melting and slowly traveling it down an ingot (cast).

Zone refining – a method of purifying a metal bar by passing it through an induction heater; this causes impurities to move along a melted portion.

Zwitterion (formerly called a dipolar ion) – a neutral molecule with a positive and negative electrical charge; multiple positive and negative charges can be present, distinct from dipoles at different locations within that molecule; known as *inner salts*.

Frank J. Addivinola, Ph.D.

This study guide's lead author and chief editor is Dr. Frank Addivinola. With his outstanding education, laboratory research, and decades of university science teaching, Dr. Addivinola lent his expertise to oversee the development of this series.

During his extensive career, Dr. Addivinola held faculty positions at colleges and universities, including Harvard University, Johns Hopkins University, University of Maryland, and Northeastern University, and taught undergraduate and graduate-level courses in biology, biochemistry, organic chemistry, inorganic chemistry, physics, anatomy and physiology, medical terminology, nutrition, and medical ethics. He received several awards for his research and presentations.

Dr. Frank Addivinola conducted original research in developmental biology as a doctoral candidate and pre-IRTA fellow in Molecular and Cell Biology at the National Institutes of Health (NIH). His dissertation advisor was Nobel laureate Marshall W. Nirenberg, Chief of the Biochemical Genetics Laboratory at the National Heart, Lung, and Blood Institute (NHLBI). Before NIH, Dr. Addivinola researched prostate cancer in the Cell Growth and Regulation Laboratory of Dr. Arthur Pardee at the Dana Farber Cancer Institute of Harvard Medical School.

Dr. Addivinola holds an undergraduate degree in biology from Williams College. He completed his Masters at Harvard University, Masters in Biotechnology at Johns Hopkins University, and five other graduate degrees at the University of Maryland University College, Suffolk University, and Northeastern University.

Electronic Structure & Periodic Table

Chemical Bonding

States of Matter & Phase Equilibria

Stoichiometry

Solution Chemistry

Chemical Kinetics & Equilibrium

Acids & Bases

Chemical Thermodynamics

Electrochemistry

Visit our Amazon store

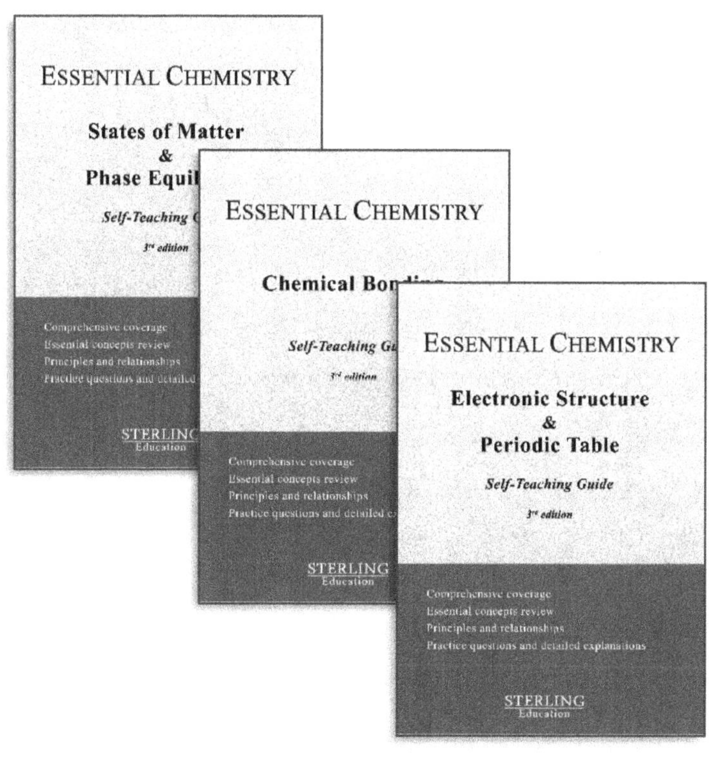

For online practice resources visit

https://www.sterling-prep.com

If you benefited from this book, please leave a review on Amazon so others
can learn from your input. Reviews help us understand our customers'
needs and experiences while keeping our commitment to quality.